Introduction to Simulation Methods for Gas Discharge Plasmas

Accuracy, reliability and limitations

Introduction to Simulation Methods for Gas Discharge Plasmas

Accuracy, reliability and limitations

Ismail Rafatov
Department of Physics, Middle East Technical University, Ankara, Turkey

Anatoly Kudryavtsev
Department of Optics, St Petersburg State University, St Petersburg, Russian Federation

IOP Publishing, Bristol, UK

ISBN 978-0-7503-2360-4 (ebook)
ISBN 978-0-7503-2358-1 (print)
ISBN 978-0-7503-2361-1 (myPrint)
ISBN 978-0-7503-2359-8 (mobi)

DOI 10.1088/978-0-7503-2360-4

Version: 20201201

IOP ebooks

British Library Cataloguing-in-Publication Data: A catalogue record for this book is available from the British Library.

Published by IOP Publishing, wholly owned by The Institute of Physics, London

IOP Publishing, Temple Circus, Temple Way, Bristol, BS1 6HG, UK

US Office: IOP Publishing, Inc., 190 North Independence Mall West, Suite 601, Philadelphia, PA 19106, USA

Contents

Preface

This book is an introduction to the numerical modeling methods for gas discharge plasmas. It is intended to assist and direct graduate students and scientists, whose research activity deals with computational plasma physics.

The fluid equations of plasma in the drift–diffusion approximation are derived from the kinetic Boltzmann equation in chapter 1. The essentials of basic modeling approaches (fluid, particle, and hybrid) for gas discharges are described in chapter 2. The details of the implementation of these methods are demonstrated by examples of glow (DC and RF) discharges. This chapter also includes the basics of the finite-difference method and a systematic description of the finite volume method for the numerical solution of the spatially one-dimensional drift–diffusion equation. Numerical investigation of non-linear dynamics and spatio-temporal pattern formation in the gas discharge system with a semiconductor cathode is presented in chapter 3.

Author biographies

Ismail Rafatov

Ismail Rafatov obtained his PhD at the KRSU (Bishkek, Kirghizia) in 1999. Since 2004 he has been working at the Department of Physics of the Middle East Technical University (Ankara, Turkey). I Rafatov is currently Professor of this department. His research interests encompass computational plasma, non-linear dynamics and chaos, and non-equilibrium pattern formation.

Anatoly Kudryavtsev

Anatoly Kudryavtsev received his MS and PhD degrees in physics from the Leningrad State University, USSR (now, Saint Petersburg State University, Russia), in 1976 and 1983, respectively. Since 1982 he has held the positions of junior researcher, senior researcher, and Associate Professor at the Department of Optics of this University. Anatoly Kudryavtsev is an expert in the physics of gas discharge plasma; he has published more than 150 papers and several monographs on the probe technique for plasma diagnostics and the physics of gas discharges.

Symbols

CCP	Capacitively coupled plasma
CFL	Courant–Friedrichs–Lewy condition
CVC	Current–voltage characteristic
DC	Direct current
DF	Distribution function
DS	Dissipative soliton
EEDF	Electron energy distribution function
FDM	Finite difference method
FVM	Finite volume method
GDP	Gas discharge plasma
GDSS	Gas discharge–semiconductor system
IEDF	Ion energy distribution function
LFA	Local field approximation
LTP	Low temperature plasma
LTE	Local thermodynamic equilibrium
λ_D	Debye length
MC	Monte Carlo
N_D	Plasma parameter
PIC/MCC	Particle in cell/Monte Carlo collision
RF	Radio frequency
Td	E/N in Townsend (1 Td = 10^{-17} V cm^2)
VDF	Velocity distribution function

IOP Publishing

Introduction to Simulation Methods for Gas Discharge Plasmas

Accuracy, reliability and limitations

Ismail Rafatov and Anatoly Kudryavtsev

Chapter 1

Modeling approaches for gas discharge plasmas

1.1 Introduction

Gas discharge plasma (GDP) is the most common type of low-temperature plasma with a large number of practical applications encompassing almost all areas of modern science and technology. There are many excellent textbooks and monographs containing detailed information on the physics and phenomenological description of various types of gas discharges. Among them are the books of Lieberman and Lichtenberg [1], Raizer [2], Chen [3], Piel [4], Smirnov [5], and others [6–9]. We should also note a number of recent reviews [10–14].

Experience of the development of gas discharge physics has shown that the number of ways of designing gas discharge devices is practically inexhaustible, which is due to the ability to effectively implement a great many different modes and regimes of GDP. This follows from the fact that the properties of GDP are determined by a large variety of processes involving charged, excited, and normal particles, their transformation into each other, interactions with the electromagnetic fields, the particle and energy transport processes, etc.

First of all, these include processes involving electrons, the behavior of which is determined by the form of their energy distribution function (EEDF). The formation of the EEDF under specific conditions is regulated by a large number of different elastic and inelastic collisions of electrons with normal and excited atoms and molecules of the background gas, as well as by the electron transport processes due to spatial gradients and external fields. Since the EEDF in the gas discharge plasma is almost invariably non-equilibrium (non-Maxwellian), the problem of finding the EEDF is one of the basic challenges when determining the gas discharge characteristics.

In turn, processes involving positive and negative atomic and molecular ions regulate the death of charges due to their bulk recombination (dissociative, radiative, triple, ion–ion, etc), attachment and detachment, and transport due to the ambipolar diffusion and drift.

doi:10.1088/978-0-7503-2360-4ch1

Although the formation of electrons and ions in the initial stage occurs due to direct ionization of neutral particles by electron impact, in the subsequent stages the ionization process is influenced by the presence of excited atoms and molecules. The latter are formed upon excitation of atoms and molecules by electron impacts and in radiative processes. Since excited particles of various states contribute to the ionization, this increases the number of regimes with different types of ionization of atoms and gas molecules. It should be mentioned that collisions involving excited atoms and molecules can also lead to ionization, as is the case for associative ionization and the Penning process.

Thus, a characteristic feature of a gas discharge plasma resides in the variety of possible modes and regimes of its existence and evolution. One of the ways to gain a detailed understanding of these processes is numerical modeling of the discharge plasma phenomena.

It is of prime importance that the problem of numerical analysis of GDP is subdivided into several subproblems, the solution to each of which requires knowledge of the solution of other subproblems, that implies the fundamentally self-consistent nature of this problem.

Calculation of electromagnetic fields. The electromagnetic field plays a dual role in the gas discharge plasma. Firstly, due to Joule heating, it forms the EEDF and creates electrons capable of ionization, which maintain the discharge. Secondly, fields in an inhomogeneous discharge plasma generate flows of charged particles and establish plasma quasi-neutrality. In a direct current (DC) discharge, an electric field performs both of these functions. In RF discharges, a high-frequency electric field is responsible for Joule heating, while particle flows and quasi-neutrality are provided by the quasi-stationary potential field. One of the best known examples is a self-consistent polarization field, providing ambipolar diffusion. The electromagnetic fields in the plasma are determined from the Maxwell equations, where the source terms are the charges and currents generated by both the external electrodes and antennas and those arising in the plasma itself.

Calculation of EEDF. If the electromagnetic fields are known, then the EEDF for electrons can be found from the solution of the Boltzmann equation. The EEDF is determined by the balance between the energy received by the electrons due to the Joule heating (diffusion in energy) and the energy loss due to collisions. Elastic collisions with neutrals is the basic mechanism of momentum relaxation, while inelastic collisions accompanied by the excitation of atoms and molecules usually lead to energy loss. The resulting EEDF, the shape of which depends on the characteristics of electron collisions, in general differs significantly from the standard Maxwell distribution. The 'tail' of the EEDF usually decreases exponentially with energy, so that its appearance is highly sensitive to specific probabilities of elementary processes. This issue is especially important when considering processes with a high energy threshold. The main process of this type is electron impact ionization, which ensures the very existence of the discharge. Calculation of the rate constants of these processes requires knowledge of the exact shape of the EEDF, otherwise, the use of the Maxwell distribution function may lead to serious uncontrolled errors.

Calculation of the plasma concentration profile. In order to determine the profiles of charged particle concentrations, it is necessary to solve a system of transport equations in which the sources and sinks of charged particles are related to their fluxes under the influence of diffusion and drift in a quasi-stationary field. For this purpose it is necessary to know the EEDF, because it defines the rates of ionization and other plasma-chemical processes and hence the birth and death of charged particles, and also the profiles of the electric and magnetic fields in the plasma.

Thus, none of these subproblems can be solved independently of the others. It is impossible to find the EEDF from the Boltzmann equation, since the field in the plasma is self-consistent and is determined not only by external electrodes, antennas, coils, etc, but also by the charges and currents of the charged plasma particles themselves. Therefore, in order to determine the fields, it is necessary to know these charges and currents profiles, which are determined by the transport equations, for the solution of which it is necessary to already know the fields. Therefore, a non-self-consistent solution to any particular problem, analytical or numerical, for example, finding the EDF for electrons subject to given fields or finding the fields arising from a given plasma configuration, have very little relation to reality.

It follows from the above that numerical solution to this self-consistent problem is inevitable. The extreme complexity of modern plasma technologies is an additional incentive for the widespread use of computer simulation methods for GDPs.

1.1.1 Basic modeling approaches

Different approaches that are used for the gas discharge modeling can by classified as:

- *fluid models*,
- *kinetic* and *particle models*, and their combinations known as
- *hybrid models*.

These models have their advantages and disadvantages. The advantages of the fluid models are their relative simplicity and computational efficiency. Fluid models are developed from the velocity moments of the kinetic Boltzmann equation for plasma particles. The resulting system of equations includes continuity, momentum, and energy balance equations for the plasma species, coupled to the Maxwell equations for the self-consistent electric and magnetic fields [2, 15, 16]. In deriving fluid equations, it is assumed that the distribution functions (DF) of all plasma species are close to Maxwellian. The corresponding criterion requires that the characteristic temporal and spatial scales of the processes, as well as of the electric and magnetic fields, exceed the temporal and spatial scales of the relaxation of the distribution functions to the Maxwellian [15]. For the elastic collisions between particles of comparable masses (such as neutral and exited particles and ions) this process occurs relatively quickly, almost within one collision time. In this case, the Maxwellian distributions of all the species with one common temperature are established. The exception is electrons whose EDF is often non-equilibrium due to their low mass.

Two main variations of the fluid model are generally used to account for the non-Maxwell EEDF. First, the *simple* fluid model, where it is assumed that the EEDF is determined by the local values of certain parameters (primarily the fields) at a given point in space and at a given instant in time. This leads to the so-called *local field approximation* (LFA), according to which the rate constants of electron-induced reactions as well as the electron transport coefficients are defined as functions of the local value of the electric field. However, in collisions of electrons with heavy particles, their energy relaxation is much slower than that of momentum relaxation and it occurs over a longer length. Therefore, in spatially inhomogeneous fields (e.g., in the near-electrode regions of discharges), the LFA is not suitable to characterize processes involving electrons and can lead to serious errors.

In order to eliminate the disadvantages associated with the LFA and incorporate to a certain extent a non-local transport of electrons into the fluid model, an approach known as the *extended* fluid model (which is also called the *local mean energy approximation*, LMEA, as opposed to LFA) was suggested in [17]. Its various modifications were applied in many studies (see, e.g., [17–26]). According to this approach, the electrons transport (mobility and diffusion) coefficients as well as the rates of the electron-induced reactions are determined as functions of the electron temperature rather than the local electric field, through the solution of the local electron Boltzmann equation. Spatial distribution of the electron temperature is obtained from the energy balance equation for electrons, which along with the volume processes includes also heat transfer by heat conduction. As a result, the profile of the electron temperature and hence profiles of the rates of the electron-induced processes are blurred by the length of the electron thermal relaxation. This implies that the electron heating source in the plasma becomes non-local.

Exact description of the electron behavior can be obtained from the solution of the kinetic Boltzmann equation [27–31]. However, this approach is very complicated mathematically, especially if more than one dimension needs to be taken into consideration. The method known as PIC/MCC (Particle in cell/Monte Carlo collision) [32, 33] which couples MC simulations for the dynamics of electrons and ions to the Poisson equation for the self-consistent electric field, is also computationally expensive, and has not found wide application in the gas discharge modeling. It is also important to note that low reliability of cross-section data for plasma processes, especially of the angular dependence of the scattering, may limit the accuracy of such detailed, *direct-method* computations. Since the reliability of a model is determined by a bottleneck, which is the least accurately known part, in solving a self-consistent problem, it is reasonable to use *equally accurate* parts. Indeed, even if methods applied to certain parts of the numerical model are sufficiently advanced and reliable, but other parts are poorly or insufficiently described, the accuracy of the results cannot be improved.

Hybrid models represent a compromise between computationally effective but approximate fluid models and accurate but time-consuming particle and kinetic models [18, 32, 34–39]. These methods are based on separating the electron ensemble into two different, independently behaving groups of low energetic (slow) and high energetic (fast) electrons. The electrons with energies above the threshold of inelastic collisions are classified as fast electrons; all others with less energy are attributed to the slow electron group. Slow (plasma) electrons, the average energy (temperature) of which is relatively small and does not exceed several eV, are essentially responsible for the electron current as well as the balance of density and temperature (mean energy) over the electron ensemble. The EDF of these electrons is close to the standard (Maxwellian), so that they are described by the fluid model with a drift–diffusion approximation for flux density. The ions and excited (metastable) particles are also described by the fluid model, due to the high charge exchange rate (for ions) and high efficiency of energy transfer in elastic collisions, they quickly come into equilibrium with a neutral gas. In turn, the characteristics of fast electrons, which are responsible for the ionization and excitation of atoms and gas molecules, are usually calculated using the Monte Carlo method. The sources of ionization and excitation thus obtained are included in the particle balance equations for slow electrons and ions, making the system self-consistent. Concentrations of fast electrons are small compared with slow electrons; their contribution to the Poisson equation can be neglected. Nowadays the hybrid models are widely used in numerical studies. Thus, this approach, based on the combination of the fluid method for ions and slow electrons and the particle/kinetic descriptions for fast electrons, was found to be one of the promising techniques for modeling the gas discharge plasma.

1.2 Boltzmann kinetic equation and derivation of balance equations for the density, momentum, and energy of plasma particles

The fluid equations for each species in plasma can be derived by taking moments of the respective kinetic Boltzmann equation,

$$\frac{\partial f_j}{\partial t} + \nabla_r \cdot \mathbf{v} f_j + \frac{q_j}{m_j} \nabla_v \cdot \left[(\mathbf{E} + \mathbf{v} \times \mathbf{B}) f_j \right] = \sum_k C_k f_k . \tag{1.1}$$

The solution $f_j(\mathbf{r}, \mathbf{v}, t)$ of this equation is a velocity distribution function defined such that $f_j d\mathbf{r} d\mathbf{v}$ represents the number of particles of species j contained in the volume $\mathbf{r} \pm d\mathbf{r}$, $\mathbf{v} \pm d\mathbf{v}$ in velocity space at time t. In this equation, $\mathbf{E}$ and $\mathbf{B}$ are the local electric and magnetic fields, m_j and q_j are the mass and the charge of the particle, ∇_r and ∇_v stand for the gradients in the coordinate and velocity spaces, respectively. The operator C_k in the right-hand side describes all (elastic and inelastic) collisional processes experienced by the particles of species k resulting in changes to f_j.

The zeroth-order moment: Let us integrate equation (1.1) over the velocity space,

$$\int \frac{\partial f_j}{\partial t} d\mathbf{v} + \int \nabla_r \cdot \mathbf{v} f_j d\mathbf{v} + \frac{q_j}{m_j} \int \nabla_v \cdot [(\mathbf{E} + \mathbf{v} \times \mathbf{B}) f_j] d\mathbf{v} = \int \left(\sum_k C_k f_k \right) d\mathbf{v}, \tag{1.2}$$

and introduce the particle density and the fluid (mass-averaged) velocity through the relations

$$n_j(\mathbf{r}, t) = \int f_j(\mathbf{r}, \mathbf{v}, t)d\mathbf{v} \tag{1.3}$$

and

$$\mathbf{u}_j(\mathbf{r}, t) = \frac{1}{n_j(\mathbf{r}, t)} \int \mathbf{v} f_j(\mathbf{r}, \mathbf{v}, t)d\mathbf{v} = \langle \mathbf{v} \rangle. \tag{1.4}$$

Observe that the third term in equation (1.2) vanishes by the Gauss theorem. The right-hand side also integrates to zero if collisions do not change the density that is the particles are neither created nor destroyed. Otherwise, the right-hand side is $\sum_k n_k \langle C_k \rangle = S_j$, and equation (1.2) obtains the form

$$\frac{\partial n_j}{\partial t} + \nabla \cdot (n_j \, \mathbf{u}_j) = S_j. \tag{1.5}$$

This is the continuity equation, which is responsible for the balance of particles of species j. The collisional term S_j in this equation accounts for all inelastic processes which impact the population of species j. The mass or charge balance equations can be obtained from (1.5) by multiplying it with the particle charge q_j or mass m_j, respectively.

The first-order moment: Next, multiplying kinetic equation (1.1) with velocity $\mathbf{v}$ and integrating over the velocity space, we obtain

$$\begin{aligned} &\int \mathbf{v}\frac{\partial f_j}{\partial t} d\mathbf{v} + \int \mathbf{v}\nabla_r \cdot \left(\mathbf{v} f_j\right) d\mathbf{v} \\ &+ \frac{q_j}{m_j} \int \mathbf{v}\nabla_v \cdot \left[(\mathbf{E} + \mathbf{v} \times \mathbf{B}) f_j\right] d\mathbf{v} = \int \mathbf{v}\left(\sum_k C_{kj} f_k\right) d\mathbf{v}. \end{aligned} \tag{1.6}$$

The first term in this equation is

$$\int \mathbf{v}\frac{\partial f_j}{\partial t} d\mathbf{v} = \frac{\partial}{\partial t}(n_j \mathbf{u}_j).$$

The second term is expressed as (see exercise 1(a))

$$\int \mathbf{v}\nabla_r \cdot \left(\mathbf{v} f_j\right) d\mathbf{v} = \nabla_r \cdot \int \mathbf{v}\mathbf{v} f_j \, d\mathbf{v} = \nabla_r \cdot n_j \langle \mathbf{v}\mathbf{v} \rangle = \nabla \cdot \left(n_j \mathbf{u}_j \mathbf{u}_j + \frac{1}{m_j} \mathbf{P}_j\right), \tag{1.7}$$

where $\mathbf{P}_j = m_j \int \mathbf{w}\mathbf{w} f_j \, d\mathbf{w} = m_j n_j \langle \mathbf{w}\mathbf{w} \rangle$ is the pressure tensor. Here, $\mathbf{w}$ denotes the random (thermal) velocity such that $\mathbf{v} = \mathbf{u}_j + \mathbf{w}$ and $\int \mathbf{w} f_j \, d\mathbf{w} = \langle \mathbf{w} \rangle = 0$. The pressure tensor $\mathbf{P}_j$ is usually expressed as $\mathbf{P}_j = p_j \mathbf{1} + \boldsymbol{\pi}_j$, where $\mathbf{1}$ is the identity tensor and $p_j = \frac{1}{3}m_j n_j \langle w^2 \rangle = \frac{1}{3}\left(P_{xx} + P_{yy} + P_{zz}\right)$ is the partial pressure, such that $\boldsymbol{\pi}_j$ is a traceless pressure tensor.

The third term in equation (1.6) (see in exercise 1(b)) can be transformed to

$$\frac{q_j}{m_j}\int \mathbf{v}\nabla_v \cdot \left[(\mathbf{E} + \mathbf{v} \times \mathbf{B})f_j\right]d\mathbf{v} = -\frac{q_j n_j}{m_j}\left(\mathbf{E} + \mathbf{u}_j \times \mathbf{B}\right). \tag{1.8}$$

We obtain eventually

$$\frac{\partial}{\partial t}(n_j\mathbf{u}_j) + \nabla \cdot (n_j\mathbf{u}_j\mathbf{u}_j) + \nabla \cdot \frac{1}{m_j}\mathbf{P}_j = \frac{q_j n_j}{m_j}(\mathbf{E} + \mathbf{u}_j \times \mathbf{B}) + \frac{1}{m_j}\mathbf{R}_j, \tag{1.9}$$

where the last term $\mathbf{R}_j/m_j = \sum_k n_k \langle C_k \mathbf{v} \rangle$ deals with the rate of momentum exchange due to collisions with different species. Further, since

$$\frac{\partial}{\partial t}(n_j\mathbf{u}_j) = \mathbf{u}_j\frac{\partial n_j}{\partial t} + n_j\frac{\partial \mathbf{u}_j}{\partial t}$$

and (see exercise 1(c))

$$\nabla \cdot (n_j\mathbf{u}_j\mathbf{u}_j) = \mathbf{u}_j\nabla \cdot (n_j\mathbf{u}_j) + n_j\mathbf{u}_j \cdot \nabla\mathbf{u}_j, \tag{1.10}$$

equation (1.9) obtains the form

$$m_j n_j\frac{d\mathbf{u}_j}{dt} = q_j n_j(\mathbf{E} + \mathbf{u}_j \times \mathbf{B}) - \nabla p_j - \nabla \cdot \boldsymbol{\pi}_j + \mathbf{R}_j, \tag{1.11}$$

which is the equation for the conservation of momentum.

The second-order moment: Finally, we derive the energy equation by multiplying kinetic equation (1.1) with v^2 and integrating over the velocity space:

$$\begin{aligned}&\int v^2\frac{\partial f_j}{\partial t}d\mathbf{v} + \int v^2\nabla_r \cdot \left(\mathbf{v}f_j\right)d\mathbf{v}\\&+\frac{q_j}{m_j}\int v^2\nabla_v \cdot \left[(\mathbf{E} + \mathbf{v} \times \mathbf{B})f_j\right]d\mathbf{v} = \int v^2\left(\sum_k C_{kj}f_k\right)d\mathbf{v}.\end{aligned} \tag{1.12}$$

This equation can be equivalently written as

$$\begin{aligned}&\frac{\partial}{\partial t}\int v^2 f_j d\mathbf{v} + \nabla_r \cdot \int v^2\mathbf{v}f_j d\mathbf{v} + \frac{q_j}{m_j}\int \nabla_v \cdot \left[v^2(\mathbf{E} + \mathbf{v} \times \mathbf{B})f_j\right]d\mathbf{v}\\&-\frac{q_j}{m_j}\int 2\mathbf{v} \cdot (\mathbf{E} + \mathbf{v} \times \mathbf{B})f_j d\mathbf{v} = \int v^2\left(\sum_k C_{kj}f_k\right)d\mathbf{v}.\end{aligned} \tag{1.13}$$

In the first term,

$$\int v^2 f_j d\mathbf{v} = n_j u_j^2 + n_j\langle w^2 \rangle.$$

In the second term (see exercise 1(d)),

$$\int v^2 \mathbf{v} f_j d\mathbf{v} = n_j u_j^2 \mathbf{u}_j + 2n_j \mathbf{u}_j \cdot \langle \mathbf{ww} \rangle + n_j \mathbf{u}_j \langle w^2 \rangle + n_j \langle w^2 \mathbf{w} \rangle \tag{1.14}$$

$$= n_j u_j^2 \mathbf{u}_j + \frac{2}{m_j} \mathbf{u}_j \cdot \mathbf{P}_j + \frac{3p_j}{m_j} \mathbf{u}_j + \frac{2}{m_j} \mathbf{q}_j, \tag{1.15}$$

where $\mathbf{q}_j = n_j \langle \frac{1}{2} m_j w^2 \mathbf{w} \rangle$ is the heat flow vector. The third term integrates to zero, while the fourth term is

$$-2\frac{q_j}{m_j} n_j \mathbf{u}_j \cdot \mathbf{E}.$$

The right-hand side of equation (1.13), multiplied with $m_j/2$ and with $\varepsilon_j = \frac{1}{2} m_j w^2$, obtains the form

$$\frac{1}{2} m_j \int v^2 \left(\sum_k C_k f_k \right) d\mathbf{v} = \mathbf{R}_j \cdot \mathbf{u}_j + \sum_k n_k \langle C_k \varepsilon_j \rangle = \mathbf{R}_j \cdot \mathbf{u}_j + Q_j,$$

which accounts for the rate at which the energy per unit volume is transferred by all other species.

For distribution function separable in real and velocity space, $f_j(\mathbf{r}, \mathbf{v}, t) = n_j(\mathbf{r}, t) F(v)$, with Maxwellian distribution function

$$F_j(v) = \left(\frac{m_j}{2\pi k_B T_j} \right)^{3/2} \exp\left(-\frac{m_j v^2}{2k_B T_j} \right),$$

defining $k_B T_j = \frac{p_j}{n_j} = \frac{1}{3} m_j \langle w^2 \rangle$, the energy equation becomes

$$\frac{\partial}{\partial t}\left(\frac{1}{2} m_j n_j u_j^2 + \frac{3}{2} n_j k_B T_j \right) + \nabla \cdot \left[\frac{1}{2} m_j n_j u_j^2 \mathbf{u}_j + \frac{5}{2} n_j k_B T_j \mathbf{u}_j + \boldsymbol{\pi}_j \cdot \mathbf{u}_j \right]$$
$$= -\nabla \cdot \mathbf{q}_j + q_j \mathbf{E} \cdot (n_j \mathbf{u}_j) + \mathbf{R}_j \cdot \mathbf{u}_j + Q_j.$$

Here, $p_j = n_j k_B T_j$ is the partial pressure, k_B is Boltzmann's constant. Note that in the first term $\frac{1}{2} n_j \left(m_j u_j^2 + 3k_B T_j \right)$ is the total random thermal and kinetic energy density. The electric field in this equation is responsible for Joule heating. The term Q_j in the right-hand side includes both elastic and inelastic collisions. By using the continuity and momentum equations (1.5) and (1.11), the energy balance equation is put into the final form (see exercise 1(e))

$$\frac{3}{2} n_j \frac{d}{dt}(k_B T_j) + p_j \nabla \cdot \mathbf{u}_j = -\boldsymbol{\pi}_j : \nabla \mathbf{u}_j - \nabla \cdot \mathbf{q}_j + Q_j. \tag{1.16}$$

Equations (1.5), (1.11), and (1.16) form the system of fluid equations for species j in plasma. Note that these equations do not form a complete set. Solution of this system requires that functions $\boldsymbol{\pi}_j$, $\mathbf{R}_j$, $\mathbf{q}_j$, and Q_j are defined in order to *close* the system.

1.3 Two-fluid equations for plasma

Let us now use the fluid equations (1.5), (1.11), and (1.16) derived in the previous section to describe the plasma of electrons and one species of positive ions, treating them as two coupled fluids. Neglecting the viscosity effects, fluid equations for the electron fraction obtain the form

$$\frac{\partial n_e}{\partial t} + \nabla \cdot n_e \mathbf{u}_e = S_e, \tag{1.17}$$

$$m_e n_e \frac{d\mathbf{u}_e}{dt} = q_e n_e (\mathbf{E} + \mathbf{u}_e \times \mathbf{B}) - \nabla p_e + \mathbf{R}_{e,i}, \tag{1.18}$$

$$\frac{3}{2} n_e \frac{dk_{\mathrm{B}} T_e}{dt} + p_e \nabla \cdot \mathbf{u}_e = -\nabla \cdot \mathbf{q}_e + Q_{e,i}. \tag{1.19}$$

Fluid equations for the ions

$$\frac{\partial n_i}{\partial t} + \nabla \cdot n_i \mathbf{u}_i = S_i, \tag{1.20}$$

$$m_i n_i \frac{d\mathbf{u}_i}{dt} = q_i n_i (\mathbf{E} + \mathbf{u}_i \times \mathbf{B}) - \nabla p_i + \mathbf{R}_{i,e}, \tag{1.21}$$

$$\frac{3}{2} n_i \frac{dk_{\mathrm{B}} T_i}{dt} + p_i \nabla \cdot \mathbf{u}_i = -\nabla \cdot \mathbf{q}_i + Q_{i,e}. \tag{1.22}$$

In these equations, n stands for the particle number density, S denotes the rate of creation (or destruction) of particles, $\mathbf{R}$ and Q denote the rates of momentum and energy exchange, respectively, m denotes the particle mass, $p = n\, k_{\mathrm{B}} T$ is the pressure, subscripts e and i indicate the electron and ion species.

Two-fluid equations for plasma need to be supplemented with Maxwell equations for the electric and magnetic fields,

$$\frac{1}{\mu_0} \nabla \times \mathbf{B} = \mathbf{j} + \varepsilon_0 \frac{\partial \mathbf{E}}{\partial t}, \tag{1.23}$$

$$\nabla \cdot \mathbf{B} = 0, \tag{1.24}$$

$$\nabla \times \mathbf{E} = -\frac{\partial \mathbf{B}}{\partial t}, \tag{1.25}$$

$$\varepsilon_0 \nabla \cdot \mathbf{E} = \sigma. \tag{1.26}$$

In these equations, σ and $\mathbf{j}$ are the charge and current densities,

$$\sigma = n_i q_i + n_e q_e, \tag{1.27}$$

$$\mathbf{j} = n_i q_i \mathbf{u}_i + n_e q_e \mathbf{u}_e, \tag{1.28}$$

where q_e and q_i are charges of the electrons and ions, respectively, ε_0 is the permittivity of free space, μ_0 is the permeability of free space.

Reasonable estimates for the friction forces $\mathbf{R}_{e,i}$ and heat transfer rates $Q_{e,i}$ in momentum and energy equations (1.18), (1.19) and (1.21), (1.22) can be obtained from the analysis of effect of collisions. If $\tau = 1/\nu_{e,i}$ is the mean free time between collisions (and hence $\nu_{e,i}$ is the collision frequency), then, since the momentum lost per collision is proportional to the relative velocity, the resulting force term can be written as

$$\mathbf{R}_{e,i} = -\nu_{e,i} m_e n_e (\mathbf{u}_e - \mathbf{u}_i) \tag{1.29}$$

and obviously $\mathbf{R}_{i,e} = -\mathbf{R}_{e,i}$. The heat transfer rate is

$$Q_{e,i} = -3\frac{m_e}{m_i}\nu_{e,i} n_e k_{\mathrm{B}}(T_e - T_i) = -Q_{i,e}. \tag{1.30}$$

Note that there are two limits where the energy equations (1.19) and (1.22) simplify. One where collisions are so rapid that $T_e \approx T_i$, and hence $Q_{e,i} = -Q_{i,e} \approx 0$. The other where collisions are very slow relative to some time-scale of interest τ,

$$\frac{1}{\tau} \gg \delta\nu_{e,i}$$

so that the heat exchange can be neglected, $Q_{e,i} = -Q_{i,e} \approx 0$. Here $\delta = 2m_e/m_i$ is the energy exchange factor. In either case energy equations (1.19) and (1.22) integrate (see exercise 1(f)) to

$$pn^{-5/3} = \mathrm{Const}, \tag{1.31}$$

the equation of state, which is the relationship between the pressure p and density n, that allows closing the system of fluid equations (1.17)–(1.22) to the particle and momentum conservation equations (1.17), (1.18) and (1.20), (1.21). In general, equation of state is $pn^{-\gamma} = \mathrm{Const}$, so that the energy equations (1.19) and (1.22) reduce to

$$p_e n_e^{-\gamma_e} = \mathrm{Const}, \tag{1.32}$$

$$p_i n_i^{-\gamma_i} = \mathrm{Const}. \tag{1.33}$$

For the isothermal compression $\gamma = 1$. With N denoting the number of degrees of freedom, $\gamma = (2 + N)/N$.

Exercise 1 (Derivation of fluid equations).

1. Derive the following equations in section 1.2:
   a) equation (1.7);
   b) equation (1.8);
   c) equation (1.10);

d) equations (1.14) and (1.15);
e) equation (1.16);
f) equation (1.31).

1.4 Fluid equations of plasma in drift–diffusion approximation

We focus on weakly ionized, low-temperature plasma of gas discharge, where the electron temperature is much higher than that of other species, that is the discharge is far from local thermodynamic equilibrium (LTE). In the corresponding discharge models, it is generally assumed that all heavy particles are characterized by the same temperature, equal to the background gas temperature and the total pressure is $p = Nk_{\mathrm{B}}T$, with the density $N = \sum_j n_j$. If the inertia terms on the left-hand side of (1.11) (and correspondingly in (1.18) and (1.21)) are ignored and $\mathbf{B} = 0$, the particle flux density for species j becomes

$$\boldsymbol{\Gamma}_j = n_j\mathbf{u}_j = \mathrm{sign}(q_j)\mu_j n_j\mathbf{E} - D_j\nabla n_j,$$

where $\mu_j = |q_j|/m_j\nu_j$ is the particle mobility, $D_j = k_BT_j\mu_j/|q_j|$ is the diffusion coefficient, and ν_j is the collision frequency.

Indeed, consider a plasma composed of the three species, namely, the electrons, positive ions, and neutral particles, which characteristics are specified below with subscripts e, i, and n. The momentum equations (1.18) and (1.21), in the absence of the magnetic field, with $\rho_{e,i}(\mathbf{r}, t) = m_{e,i}n_{e,i}(\mathbf{r}, t)$ denoting the mass density, obtain the form

$$\rho_e\frac{\partial\mathbf{u}_e}{\partial t} + \rho_e(\mathbf{u}_e\cdot\nabla)\mathbf{u}_e = -\nabla p_e - en_e\mathbf{E} - m_en_e\nu_{e,n}(\mathbf{u}_e - \mathbf{u}_n) - m_en_e\nu_{e,i}(\mathbf{u}_e - \mathbf{u}_i), \quad (1.34)$$

$$\rho_i\frac{\partial\mathbf{u}_i}{\partial t} + \rho_i(\mathbf{u}_i\cdot\nabla)\mathbf{u}_i = -\nabla p_i + en_i\mathbf{E} - m_in_i\nu_{i,e}(\mathbf{u}_i - \mathbf{u}_e) - m_in_i\nu_{i,n}(\mathbf{u}_i - \mathbf{u}_n). \quad (1.35)$$

In these equations, $\mathbf{u}_n$ denotes the mass-average velocity of the neutral gas atoms, $q_e = -e$ and $q_i = e$. In view of the fact that $m_e \ll m_i$, it is reasonable to assume $\rho_e(\mathbf{u}_e\cdot\nabla)\mathbf{u}_e \ll \rho_i(\mathbf{u}_i\cdot\nabla)\mathbf{u}_i$. In this case, equation (1.34) can be simplified to [40]

$$-\nabla p_e - en_e\mathbf{E} - m_en_e\nu_{e,n}(\mathbf{u}_e - \mathbf{u}_n) - m_en_e\nu_{e,i}(\mathbf{u}_e - \mathbf{u}_i) = 0. \quad (1.36)$$

Since $\mathbf{u}_e \gg \mathbf{u}_n$, $\mathbf{u}_i$ and $p_e = n_ek_{\mathrm{B}}T_e$, this equation can be further reduced to

$$k_{\mathrm{B}}T_e\nabla n_e + en_e\mathbf{E} + (m_e\nu_e)n_e\mathbf{u}_e = 0, \quad (1.37)$$

and finally

$$\boldsymbol{\Gamma}_e = n_e\mathbf{u}_e = -\mu_en_e\mathbf{E} - D_e\nabla n_e, \quad (1.38)$$

where $\mu_e = e/(m_e\nu_e)$ is the electron mobility and $D_e = k_{\mathrm{B}}T_e\mu_e/e$ is the electron diffusion coefficient, $\nu_e = \nu_{e,n} + \nu_{e,i}$ denotes the electron collision frequency.

Let us now consider the momentum conservation equation for the ionic species (1.35), which can be written in the form [40]

$$-\nabla p_i + en_i\mathbf{E} - m_i n_i \nu_{i,e}(\mathbf{u}_i - \mathbf{u}_e) - m_i n_i \nu_{i,n}(\mathbf{u}_i - \mathbf{u}_n) = 0. \tag{1.39}$$

Taking into account that

$$\nu_{i,e} m_i n_i(\mathbf{u}_i - \mathbf{u}_e) = -\nu_{e,i} m_e n_e(\mathbf{u}_e - \mathbf{u}_i),$$

neglecting $\mathbf{u}_n$, and since $m_e \nu_e \ll m_i \nu_{i,n}$, equation (1.39) is reduced to

$$\Gamma_i = n_i \mathbf{u}_i = \mu_i n_i \mathbf{E} - D_i \nabla n_i, \tag{1.40}$$

where $\mu_i = e/m_i \nu_{i,n}$ is the ion mobility and $D_i = k_B T_i \mu_i / e$ is the ion diffusion coefficient. When the expressions for the electron and ion flux densities $n_e \mathbf{u}_e$ and $n_i \mathbf{u}_i$ from (1.38) and (1.40) are substituted into the corresponding continuity equations (1.17) and (1.20), then the two-fluid equations with drift–diffusion approximation for the particle flux densities of the form

$$\frac{\partial n_j}{\partial t} + \nabla \cdot \left(\mathrm{sign}(q_j)\mu_j n_j \mathbf{E} - D_j \nabla n_j\right) = S_j, \tag{1.41}$$

where the subscript j indicates the species, are obtained. The right-hand side term S_j describe the rate at which species j is created or destroyed in collision processes. In maintaining the lower-pressure discharges, excitation and ionization of atoms by electrons are the dominating inelastic processes. In order to implement a self-consistent description, particle balance equations (1.41) are supplied with Poisson equation for electric field. Models based on these equations are considered in the following sections.

In practice, gas discharges are essentially maintained by electron impact ionization in an external electric field. Since the rates of collisional processes are proportional to the total particle density, their efficiency is characterized by the reduced electric field, E/N, the ratio of local value of the electric field E to the total particle density $N = \sum_j n_j$, which is usually measured in Townsend (1 Td = 10^{-21} V m^2). The density N in this relation can be replaced with the neutral particle density that provides reasonable accuracy. Alternatively, the ratio E/p of the local field to the total particle pressure p, which is commonly assumed to be constant within the discharge, is utilized. The gas discharge models where the particle transport (mobility and diffusion) coefficients as well as the particle creation/destruction rates are defined as functions of the reduced electric field, E/N, are known as the *local field approximation* (LFA) models. According to this approximation, the electrons are considered to be in equilibrium with the local electric field, that is the local electron energy distribution function (EEDF) depends only on the local value of the electric field.

1.5 Limitations and applicability of the fluid model

The main criterion for the applicability of the fluid model for particles of type j is the small mean free path λ_j of this type of particles in comparison with the characteristic size L of the system (see, e.g., [15] for details)

$$L > \lambda_j. \tag{1.42}$$

It should be mentioned that the fluid model is not applicable at distances from the plasma boundaries that are smaller than the particle mean free path. However, the results of the kinetic analysis show that the formulation of effective boundary conditions is sufficient to complete the equations of the fluid model.

For particles of comparable masses (such as neutral and excited atoms and molecules and their ions), the momentum and energy relaxations occur in one (several) collision. Therefore, when the condition (1.42) is satisfied, the EDFs of neutral and excited particles are close to Maxwellian with room (or thermostat) temperature.

Unlike neutral gas particles, charged particles in plasma are affected by an electric field (in the general case, by an electromagnetic field), which heats them. In this situation, their temperatures (mean energies) exceed the temperature of the thermostat, and their EDFs are close to Maxwellian only at sufficiently high degrees of ionization (greater than 10^{-4}–10^{-3}), when the relaxation of their energies is determined by electron–electron and ion–ion collisions.

EDF of ions in a strong electric field, in the case of low ionized plasma (with a degree of ionization less than 10^{-4}–10^{-3}) is non-Maxwellian. However, in the relevant fluid equations, the characteristics of ions contribute only to the definitions of their transport (mobility and diffusion) coefficients, which are integral characteristics of the entire ensemble of these particles and are not sensitive to the actual shape of their EDF. Therefore, the characteristics of ions can be approximately calculated with the Maxwellian EDF, but with the temperature determined from the energy balance taking account of heating in an electric field [2, 15].

For electrons, which are the most mobile particles and whose mass is small compared to the masses of neutral particles and ions, the characteristic scales of momentum and energy relaxations are significantly different. Indeed, for collisions between electrons and heavy particles, the energy relaxation length λ_ε and time τ_ε considerably exceed the momentum relaxation length and time, which are equal, respectively, to the electron mean free path λ and the inverse of the momentum transfer collision frequency $1/\nu_{e,a}$. In fact, for elastic collisions (see, e.g., [15] for details)

$$\lambda_\varepsilon = \lambda/\sqrt{\delta} > 100\,\lambda, \tag{1.43}$$

$$\tau_\varepsilon = 1/\delta\nu_{e,a} > 100^2/\nu_{e,a}, \tag{1.44}$$

where $\delta = 2m_e/m_a$ is the energy exchange factor.

In a weak electric field, where the energy gained by an electron over the energy relaxation length is small compared to the temperature of heavy particles, the electron distribution function is close to Maxwellian with a temperature equal to the temperature of heavy particles. In strong fields, the situation depends on the degree of ionization.

At a relatively high degree of ionization, such that the frequency of electron–electron collisions $\nu_{e,e} > \delta\,\nu_{e,a}$, the EEDF is close to Maxwellian with the electron partial temperature, and introduction of the set of three balance equations (for the electron density, momentum, and energy) is reasonable. They are derived, if the scales of the electric field variation exceed the time and length scales of Maxwellization, caused by the electron–electron collisions. This approximation corresponds to *the standard fluid approach* [15] (see also sections 1.3 and 1.4).

In the opposite case of lower ionization degree, the situation is much more complicated, it depends on factors such as the degree of ionization, the characteristics of collisions and the strength of the electric field. If the degree of ionization is very low, such that $\nu_{e,e} < \delta\, \nu_{e,a}$, the distribution deviates significantly from the Maxwellian in strong electric fields, and the kinetic description for electrons should be used.

The rate at which the charged particles temperature (mean energy) relaxes in the electric field, is determined by the relevant time τ_ε and length λ_ε scales of energy relaxation. If variation in the electric field over the time τ_ε is relatively weak, the mean energy (and correspondingly the EDF) responds to (follows) the changes in the field (which can be considered as quasi-stationary), and not if the changes are fast. A relevant criterion for the electric field can be formulated by the relation

$$\left(\frac{1}{E}\frac{dE}{dt}\right)\tau_\varepsilon \ll 1. \tag{1.45}$$

In turn, a field can be considered as uniform if it changes slowly over a distance of the order of the energy relaxation length λ_ε, i.e., under the condition

$$\left(\frac{1}{E}\frac{dE}{dx}\right)\lambda_\varepsilon \ll 1. \tag{1.46}$$

In conditions (1.45) and (1.46), the energy relaxation time and length scales are determined by the collision frequency and mean free path, for electrons from equations (1.43) and (1.44), and for ions these are defined by the relations $\tau_\varepsilon = 1/\nu_{i,a}$ and $\lambda_\varepsilon = \lambda_{i,a}$.

In the case of strongly inhomogeneous fields, where the relations (1.45) and (1.46) are not satisfied, the distribution function and temperature (mean energy) are no longer functions of the local value of the field at a given point in space. In fact, they are determined by the region of size of the order λ_ε, and, accordingly, by the time interval of the order τ_ε. This behavior is usually referred to as non-localitiy of EDF. In this situation, the LFA is no longer valid, and relevant plasma species (usually electrons) must be treated kinetically.

Schematic diagram in figure 1.1 shows regions in the parameter space, spanned by the characteristic plasma size L and gas pressure p, where different modeling approaches, namely, *kinetic*, *hybrid*, and *fluid*, are applicable. As can be seen from this figure, the non-local kinetic effects become important in the parameter regimes where the characteristic system size L is smaller than the energy relaxation length λ_ε. For example, fluid model for the electronic component becomes inapplicable in the case of short discharges in terms of the product pL in the region where $L < \lambda_\varepsilon$. It

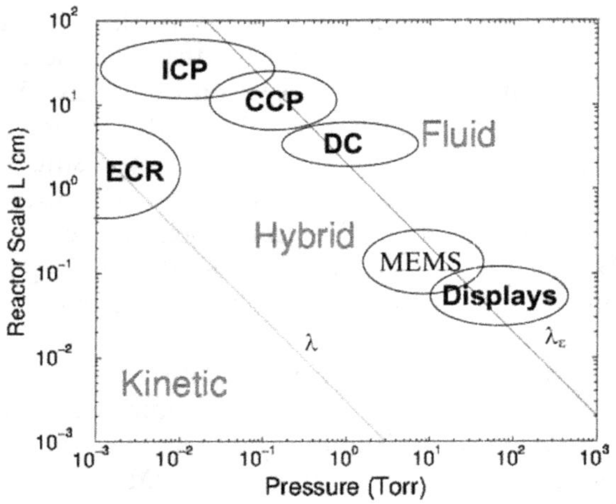

Figure 1.1. Schematic diagram of the parameter ranges, where different models (kinetic, hybrid, and fluid) can be applied. Adapted from [41], copyright (2003), with permission of Elsevier.

should be mentioned that according to analysis in [42], plasma of atomic gases is non-local in the range $pL < 10$ cm Torr, which involves low and medium pressure gas discharges, and also microdischarges at high pressures.

Thus, condition (1.42) is the main criterion for the validity of using the fluid model for all components of the system.

References

[1] Lieberman M A and Lichtenberg A J 2005 *Principles of Plasma Discharges and Materials Processing* (New York: Wiley)

[2] Raizer Y P 1991 *Gas Discharge Physics* (Berlin: Springer)

[3] Chen F F 2016 *Introduction to Plasma Physics and Controlled Fusion* (Berlin: Springer)

[4] Piel A 2017 *Plasma Physics: An Introduction to Laboratory, Space, and Fusion Plasmas* (Berlin: Springer)

[5] Smirnov B M *et al* 2015 *Theory of Gas Discharge Plasma* (Berlin: Springer)

[6] Roth J R 1995 *Industrial Plasma Engineering* vol. 1 (Bristol, UK: Institute of Physics Publishing)

[7] Meichsner J, Schmidt M, Schneider R and Wagner H-E 2012 *Nonthermal Plasma Chemistry and Physics* (Boca Raton, FL: CRC Press)

[8] Becker K H, Kogelschatz U, Schoenbach K H and Barker R J 2004 *Non-equilibrium Air Plasmas at Atmospheric Pressure* (Boca Raton, FL: CRC Press)

[9] Fridman A 2008 *Plasma Chemistry* (Cambridge: Cambridge University Press)

[10] Gudmundsson J T and Hecimovic A 2017 Foundations of dc plasma sources *Plasma Sources Sci. Technol.* **26** 123001

[11] Von Keudell A and Schulz-Von Der Gathen V 2017 Foundations of low-temperature plasma physics—an introduction *Plasma Sources Sci. Technol.* **26** 113001

[12] Murphy A B and Uhrlandt D 2018 Foundations of high-pressure thermal plasmas *Plasma Sources Sci. Technol.* **27** 063001

[13] Oehrlein G S and Hamaguchi S 2018 Foundations of low-temperature plasma enhanced materials synthesis and etching *Plasma Sources Sci. Technol.* **27** 023001

[14] Bruggeman P J, Iza F and Brandenburg R 2017 Foundations of atmospheric pressure non-equilibrium plasmas *Plasma Sources Sci. Technol.* **26** 123002

[15] Rozhansky V A and Tsendin L D 2001 *Transport Phenomena in Partially Ionized Plasma* (Boca Raton, FL: CRC Press)

[16] Boeuf J-P 1987 Numerical model of RF glow discharges *Phys. Rev.* A **36** 2782

[17] Boeuf J P and Pitchford L C 1995 Two-dimensional model of a capacitively coupled RF discharge and comparisons with experiments in the gaseous electronics conference reference reactor *Phys. Rev.* E **51** 1376

[18] Derzsi A, Hartmann P, Korolov I, Karacsony J, Bánó G and Donkó Z 2009 On the accuracy and limitations of fluid models of the cathode region of dc glow discharges *J. Phys. D: Appl. Phys.* **42** 225204

[19] Arslanbekov R R and Kolobov V I 2003 Two-dimensional simulations of the transition from Townsend to glow discharge and subnormal oscillations *J. Phys. D: Appl. Phys.* **36** 2986

[20] Grubert G K, Becker M M and Loffhagen D 2009 Why the local-mean-energy approximation should be used in hydrodynamic plasma descriptions instead of the local-field approximation *Phys. Rev.* E **80** 036405

[21] Hagelaar G J M and Pitchford L C 2005 Solving the Boltzmann equation to obtain electron transport coefficients and rate coefficients for fluid models *Plasma Sources Sci. Technol.* **14** 722

[22] Sakiyama Y, Graves D B and Stoffels E 2008 Influence of electrical properties of treated surface on RF-excited plasma needle at atmospheric pressure *J. Phys. D: Appl. Phys.* **41** 095204

[23] Bogdanov E A, Adams S F, Demidov V I, Kudryavtsev A A and Williamson J M 2010 Influence of the transverse dimension on the structure and properties of dc glow discharges *Phys. Plasmas* **17** 103502

[24] Bogdanov E A, Kapustin K D, Kudryavtsev A A and Chirtsov A S 2010 Different approaches to fluid simulation of the longitudinal structure of the atmospheric-pressure microdischarge in helium *Tech. Phys.* **55** 1430–42

[25] Becker M M, Loffhagen D and Schmidt W 2009 A stabilized finite element method for modeling of gas discharges *Comput. Phys. Commun.* **180** 1230–41

[26] Rafatov I, Bogdanov E A and Kudryavtsev A A 2012 On the accuracy and reliability of different fluid models of the direct current glow discharge *Phys. Plasmas* **19** 033502

[27] Winkler R, Arndt S, Loffhagen D, Sigeneger F and Uhrlandt D 2004 Progress of the electron kinetics in spatial and spatiotemporal plasma structures *Contrib. Plasma Phys* **44** 437–49

[28] Kolobov V I and Arslanbekov R R 2006 Simulation of electron kinetics in gas discharges *IEEE Trans. Plasma Sci.* **34** 895–909

[29] Robson R E, White R D and Lj Petrović Z 2005 Colloquium: physically based fluid modeling of collisionally dominated low-temperature plasmas *Rev. Mod. Phys.* **77** 1303

[30] White R D, Robson R E, Dujko S, Nicoletopoulos P and Li B 2009 Recent advances in the application of Boltzmann equation and fluid equation methods to charged particle transport in non-equilibrium plasmas *J. Phys. D: Appl. Phys.* **42** 194001

[31] Yuan C, Bogdanov E A, Eliseev S I and Kudryavtsev A A 2017 1D kinetic simulations of a short glow discharge in helium *Phys. Plasmas* **24** 073507

[32] Donkó Z, Hartmann P and Kutasi K 2006 On the reliability of low-pressure dc glow discharge modelling *Plasma Sources Sci. Technol.* **15** 178

[33] Donkó Z 2011 Particle simulation methods for studies of low-pressure plasma sources *Plasma Sources Sci. Technol.* **20** 024001

[34] Fiala A, Pitchford L C and Boeuf J P 1994 Two-dimensional, hybrid model of low-pressure glow discharges *Phys. Rev.* E **49** 5607

[35] Boeuf J-P and Marode E A 1982 Monte Carlo analysis of an electron swarm in a nonuniform field: the cathode region of a glow discharge in helium *J. Phys. D: Appl. Phys.* **15** 2169

[36] Surendra M, Graves D B and Jellum G M 1990 Self-consistent model of a direct-current glow discharge: treatment of fast electrons *Phys. Rev.* A **41** 1112

[37] Kushner M J 2009 Hybrid modelling of low temperature plasmas for fundamental investigations and equipment design *J. Phys. D: Appl. Phys.* **42** 194013

[38] Van Dijk J, Kroesen G M W and Bogaerts A 2009 Plasma modelling and numerical simulation *J. Phys. D: Appl. Phys.* **42** 190301

[39] Bogaerts A, Gijbels R and Goedheer W J 1995 Hybrid Monte Carlo-fluid model of a direct current glow discharge *J. Appl. Phys.* **78** 2233–41

[40] Surzhikov S and Shang J 2014 Normal glow discharge in axial magnetic field *Plasma Sources Sci. Technol.* **23** 054017

[41] Kolobov V and Arslanbekov R 2003 *Microelectron. Eng.* **69** 606

[42] Tsendin L D 1995 Electron kinetics in non-uniform glow discharge plasmas *Plasma Sources Sci. Technol.* **4** 200

Chapter 2

Numerical simulation of gas discharges: fluid, particle, and hybrid methods

The modeling approaches that are generally used in gas discharge studies can be classified as fluid methods, kinetic/particle methods and their combinations, known as hybrid methods, which represent a compromise between the computationally very efficient fluid models and fully kinetic particle models that require very extensive computations.

In this chapter, the basic concepts and definitions which are needed for the numerical solution of the drift–diffusion equation are introduced in section 2.1. Then, in section 2.2, the finite volume method (FVM) for the solution of this equation is described systematically. *Simple* and *extended* fluid models are introduced and illustrated by the example of direct current (DC) glow discharge in section 2.3. Finally, the Particle in cell/Monte Carlo collision (PIC/MCC) and hybrid Monte Carlo–fluid models are discussed in sections 2.4 and 2.5 and demonstrated by using examples of DC glow discharge and radio frequency capacitively coupled plasma (RF CCP) discharge.

2.1 Preliminary technique

As we have seen in section 1.4 (see equation (1.41)), with the drift–diffusion approximation for particle fluxes, continuity equation for species of type k in plasma obtains the form

$$\frac{\partial n_k}{\partial t} + \nabla \cdot \left(s_k \mu_k n_k \mathbf{E} - D_k \nabla n_k\right) = S_k \tag{2.1}$$

or

$$\frac{\partial n_k}{\partial t} + \nabla \cdot \mathbf{\Gamma}_k = S_k \tag{2.2}$$

doi:10.1088/978-0-7503-2360-4ch2

with $\mathbf{\Gamma}_k = s_k \mu_k n_k \mathbf{E} - D_k \nabla n_k$ denoting the total flux density, consisting of drift and diffusion components. In this equation, n_k is the particle number density, μ_k and D_k are the mobility and diffusion coefficients, $s_k = \text{sign}(q_k)$ is either plus or minus depending on the charge q_k, $\mathbf{E}$ is the electric field strength, S_k is the source term, which is determined by the processes of creation and destruction (such as ionization, recombination, etc) of particles of type k.

In the description of numerical models, we will restrict our consideration to the case of one space dimension x. In this situation, omitting subscripts and denoting the drift velocity as $v = \text{sign}(q)\mu E$, equation (2.1) obtains the form

$$\frac{\partial n}{\partial t} + \frac{\partial}{\partial x}\left(vn - D\frac{\partial n}{\partial x}\right) = S, \tag{2.3}$$

where the total flux density is $\Gamma = vn - D\frac{\partial n}{\partial x}$. In the following, first, the basic concepts and definitions of computational methods, related to the solution of this equation, are introduced. Next, the basics of FVM for solving equation (2.3) are considered systematically.

2.1.1 Basic concepts and definitions

Grid functions

Consider a grid Ω^h in the one-dimensional domain $\Omega = [c, d]$ defined by

$$\Omega^h = \left\{ x_i \;\middle|\; x_i \equiv c + (i-1)h, \quad h = \frac{d-c}{N-1}, \quad i = 1, 2, \dots, N \right\}.$$

Here x_i $(i = 1, 2, \dots, N)$ are grid points, h is a step size of the grid. Since

$$x_{i+1} - x_i = h$$

is constant for all $(i = 1, 2, \dots, N-1)$, this is a uniform (equidistant) grid.

A function defined on the points of the grid Ω^h is called a grid function and denoted by

$$f^h = \left(f_1^h, f_2^h, \dots, f_N^h\right).$$

The magnitudes of grid functions are expressed by their norms, which can be defined by the formula

$$\|f^h\|_p = \left(\frac{1}{N}\sum_{i=1}^{N} |f_i^h|^p\right)^{\frac{1}{p}}, \tag{2.4}$$

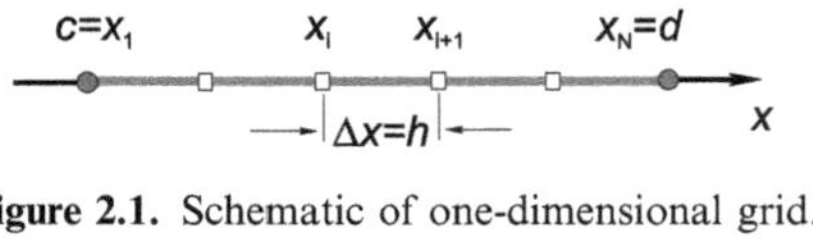

Figure 2.1. Schematic of one-dimensional grid.

where p is a positive integer. Frequently used is a *maximum norm* defined by

$$\|f^h\|_\infty = \max_{1\leqslant i\leqslant N} |f_i^h|.$$

In order to be able to compare continuous and grid functions, the projection operation, denoted by a symbol $(.)^h$, is introduced. Projection of a continuous function $f(x)$ onto grid Ω^h is a grid function defined by the formula

$$(f(x))^h = (f(x_1), f(x_2), \ldots, f(x_N)).$$

Note that a continuous function $f(x)$ defined on the interval $\Omega = [c, d]$ can be specified by the norm of the form similar to that in (2.4):

$$\|f(x)\|_p = \left(\frac{1}{d-c}\int_c^d |f(x)|^p dx\right)^{\frac{1}{p}}, \tag{2.5}$$

where p is a positive integer. The maximum norm is defined correspondingly by

$$\|f(x)\|_\infty = \max_{c\leqslant x\leqslant d} |f(x)|.$$

Approximation of differential operator by difference operators

Let $u^h = (u_1^h, u_2^h, \ldots, u_N^h)$ be a grid function defined on the grid Ω^h of equidistant points x_i $(i = 1, 2, \ldots, N)$ with step size h. The grid operator L^h acts as

$$L^h u_i^h = \sum_{j=1}^{N} a_{i,j} u_j^h \quad (i, j = 1, 2, \ldots, N)$$

so that it can be defined by $N \times N$ square matrix

$$L^h = [a_{i,j}]_{i,\, j=1}^N.$$

Consider the following grid operators approximating derivatives of function $u(x)$ at grid point x_i:

$$u'(x_i) \approx \frac{u(x_{i+1}) - u(x_i)}{h} \equiv D_+ u(x_i) \quad (i = 1, 2, \ldots, N-1) \tag{2.6}$$

known as *a forward difference* operator,

$$u'(x_i) \approx \frac{u(x_i) - u(x_{i-1})}{h} \equiv D_- u(x_i) \quad (i = 2, 3, \ldots, N) \tag{2.7}$$

which is *a backward difference* operator, and

$$u'(x_i) \approx \frac{u(x_{i+1}) - u(x_{i-1})}{2h} \equiv D_0 u(x_i) \quad (i = 2, 3, \ldots, N-1) \tag{2.8}$$

which is called *a central difference* operator. Applying operators D_+ and D_- successively one after another, we obtain an approximation for the second order derivative

$$u''(x_i) \approx \frac{u(x_{i+1}) - 2u(x_i) + u(x_{i-1})}{h^2} \equiv D_+D_-u(x_i) \quad (i = 2, 3, \dots, N-1) \qquad (2.9)$$

Exercise 2 (Finite-difference operators)

1. Derive the following estimates of accuracy for the finite-difference operators:
   a) $\|D_+(u(x))^h - (u'(x))^h\|_\infty \leqslant Ch$, where $C = \frac{1}{2}\left\|u''(x)\right\|_\infty$.
   b) $\|D_-(u(x))^h - (u'(x))^h\|_\infty \leqslant Ch$, where $C = \frac{1}{2}\left\|u''(x)\right\|_\infty$.
   c) $\|D_0(u(x))^h - (u'(x))^h\|_\infty \leqslant Ch^2$, where $C = \frac{1}{6}\left\|u'''(x)\right\|_\infty$.
   d) $\|D_+D_-(u(x))^h - (u''(x))^h\|_\infty \leqslant Ch^2$, where $C = \frac{1}{12}\left\|u^{(\mathrm{iv})}(x)\right\|_\infty$.

Concepts of consistency, stability, and convergence

Let Ω^h be a uniform grid with step size h introduced in the interval Ω of the real line. Let function $u = u(x)$ defined on the interval Ω be a solution of the differential (continuous) problem

$$Lu(x) = f(x) \quad (x \in \Omega) \qquad (2.10)$$

with differential operator L, and let the function $u^h = (u_1^h, u_2^h, \dots, u_N^h)$ defined on the grid Ω^h be the solution of the difference (discrete) problem

$$L^h u^h = f^h \qquad (2.11)$$

with grid operator (difference operator) L^h. We will consider the operator L^h of the problem (2.11) and its solution u^h as approximating the operator L and solution $u(x)$ of the continuous problem (2.10) on the grid Ω^h, respectively. Then, the truncation error T^h is a grid function defined by

$$T^h = (Lu)^h - L^h(u)^h, \qquad (2.12)$$

where the brackets mean projection onto the grid Ω^h. The difference between the exact solution $u(x)$ of the differential problem (2.10) and the exact solution u^h of the difference problem (2.11) is called the solution error and is denoted by $e_i^h = u(x_i) - u_i^h$ $(i = 1, 2, \dots, N)$. That is the total error in the solution is expressed by the grid function

$$e^h = (u)^h - u^h. \qquad (2.13)$$

Consistency: A difference operator L^h is said to be consistent with a differential operator L, if

$$\left\|T^h\right\|_p = \left\|(Lu)^h - L^h(u)^h\right\|_p \underset{h\to 0}{\longrightarrow} 0.$$

Further, if there exist numbers $C > 0$ and $\alpha > 0$ independent on h such that

$$\left\|T^h\right\|_p = \left\|(Lu)^h - L^h(u)^h\right\|_p \leqslant Ch^\alpha, \tag{2.14}$$

then we say that the order of approximation of operator L by difference operator L^h is α.

A difference (discrete) problem (2.11), which approximates an original differential (continuous) equation (2.10), is said to be consistent with it, if equation (2.11) becomes equivalent to (2.10) as the grid spacing tends to zero, $h \to 0$. Thus, a difference problem (2.11) is consistent with a differential problem (2.10), if

$$\left\|L^h(u)^h - f^h\right\|_p \underset{h\to 0}{\longrightarrow} 0.$$

Further, if there exist numbers $C > 0$ and $\alpha > 0$ independent on h such that

$$\left\|L^h(u)^h - f^h\right\|_p \leqslant Ch^\alpha, \tag{2.15}$$

then we say that the order of approximation of the differential problem (2.10) by the difference problem (2.11) is α.

Example 2.1.1. Let us show that the backward difference operator $L^h = D_-$ is consistent with the differentiation operator $L = d/dx$, and the order of approximation is of first-order in step size h. In fact, this question is equivalent to that in exercise 2(b). Indeed, in the maximum norm,

$$\begin{aligned}\|(Lu)^h - L^h(u)\|_\infty &= \|(u')^h - D_-(u)^h\|_\infty = \max_{2\leqslant i\leqslant N} |\, u'(x_i) - D_-u(x_i)\,| \\ &\leqslant \frac{h}{2}\, \|u''(x)\|_\infty = Ch.\end{aligned}$$

Example 2.1.2. Let us show that the discrete problem

$$\begin{cases} L^h u_i^h \equiv D_- u_i^h = f(x_i) = f_i^h, & i = 2, \dots, N \\ L^h u_1^h \equiv u_1^h = \varphi \end{cases}$$

is consistent to the differential problem

$$\begin{cases} Lu(x) \equiv u'(x) = f(x), & x \in (0, 1] \\ Lu(0) \equiv u(0) = \varphi, \end{cases}$$

and that the order of approximation is of first-order in step size h. Indeed, in the maximum norm,

$$\begin{aligned}
\|L^h(u)^h - f^h\|_\infty &= \max_{2\leqslant i\leqslant N} \left| D_{-}u(x_i) - f_i^h \right| \\
&= \max_{2\leqslant i\leqslant N} \left| (D_{-}u(x_i) - u'(x_i)) + (u'(x_i) - f(x_i)) \right| \\
&= \max_{2\leqslant i\leqslant N} \left| D_{-}u(x_i) - u'(x_i)\right| + \max_{2\leqslant i\leqslant N} \left| u'(x_i) - f(x_i)\right| \leqslant Ch,
\end{aligned}$$

where we used the estimate from exercise 2(b) and that $u'(x_i) - f(x_i) = 0$ at all points x_i.

Convergence: A solution u^h of the difference problem (2.11), which approximates a differential equation (2.10), is said to be convergent if the approximate solution u^h approaches the exact solution $u(x)$ of the differential equation as the grid spacing tends to zero, $h \to 0$. Thus, we say that the solution of the difference problem (2.11) converges to the solution of the differential problem (2.10) if

$$\|e^h\|_p = \|(u)^h - u^h\|_p \xrightarrow[h\to 0]{} 0.$$

Further, if there exist numbers $C > 0$ and $\alpha > 0$ independent on h such that

$$\left\|e^h\right\|_p = \left\|(u)^h - u^h\right\|_p \leqslant Ch^\alpha,$$

then we say that the order of convergence of the solution u^h of the discrete problem (2.11) to the solution $u(x)$ of the continuous problem (2.10) is α.

Stability: The concept of stability is concerned with the tendency of errors (such as spontaneous perturbations induced by round-off errors) accumulating in the course of computation to grow or decay. If for any right-hand side function f^h

1. solution of the discrete problem (2.11) exists uniquely, and also
2. there exists a constant $C > 0$ independent on h such that

$$\left\|u^h\right\|_p \leqslant C \left\|f^h\right\|_p \tag{2.16}$$

then the discrete problem (2.11) is said to be well-posed.

The second of these conditions ensures continuous and uniform in h dependence of the solution u^h of the problem (2.11) on the right-hand side function f^h, and hence it provides the stability of the discrete problem.

Assume that the differential problem (2.10) is well-posed and that the discrete problem (2.11) is also well-posed and consistent with the differential problem (2.10). Then, according to the Lax equivalence theorem, solution of the discrete problem converges to the solution of the differential problem, and convergence order is equal to the approximation order.

The Lax equivalence theorem: *Given a well-posed initial-value problem and its finite-difference approximation that satisfies the consistency condition, then stability is the necessary and sufficient condition for convergence.*

Schematically, this theorem can be expressed as

$$\textit{Stability} + \textit{Consistency} \Rightarrow \textit{Convergence}.$$

Proof: Let us denote $g^h = L^h(u)^h - f^h$. Then,

$$L^h(u)^h - f^h = L^h(u)^h - L^h u^h = L^h[(u)^h - u^h],$$

and hence the error in the solution $e^h = (u)^h - u^h$ is a solution of the problem

$$L^h e^h = g^h.$$

Then, by the stability condition (2.16) and consistency estimate (2.15),

$$\left\|e^h\right\|_p \leqslant C_1 \left\|g^h\right\|_p = C_1 \left\|L^h(u)^h - f^h\right\|_p \leqslant C_1 C_2\, h^\alpha = C h^\alpha,$$

and hence

$$\left\|(u)^h - u^h\right\|_p \leqslant C h^\alpha.$$

Exercise 3 (Consistency, stability, convergence)

1. Find the order of approximation of the differential operators by the finite-difference counterparts:
   a) $L = d/dx$ and $L^h = D_-$
   b) $L = d/dx$ and $L^h = D_+$
   c) $L = d/dx$ and $L^h = D_0$
   d) $L = d^2/dx^2$ and $L^h = D_+D_-$
2. Show that the discrete problem

$$\begin{cases} L^h u_i^h \equiv D_- u_i^h = f_i^h, & i = 2, \dots, N \\ L^h u_1^h \equiv u_1^h = \varphi. \end{cases}$$

   is well-posed.
3. Derive finite-difference problem approximating the differential problem

$$\begin{cases} Lu(x) \equiv u''(x) = f(x), & x \in (0, 1) \\ Lu(0) \equiv u(0) = \varphi, \\ Lu(1) \equiv u(1) = \psi. \end{cases}$$

Find the order of approximation of this problem by its finite-difference counterpart.

2.1.2 Finite-difference schemes for steady convection–diffusion equation

In order to introduce the basic concepts of finite-difference methods, we consider the second order linear differential equation

$$-\varepsilon u''(x) - a(x)u'(x) + b(x)u(x) = f(x), \quad x \in (c, d) \tag{2.17}$$

subject to the boundary conditions

$$\xi_0\, u(c) - \eta_0 \varepsilon u'(c) = \varphi, \tag{2.18}$$

and

$$\xi_1 \, u(d) + \eta_1 \varepsilon u'(d) = \psi. \tag{2.19}$$

Here, a, b, and f are functions defined in the interval $c \leqslant x \leqslant d$, ε is a positive constant,

$$\xi_0, \xi_1, \eta_0, \eta_1 \geqslant 0, \quad \xi_0 + \eta_0 > 0, \quad \xi_1 + \eta_1 > 0.$$

When formulated in this general form, the problem is called the Robin boundary-value problem. If $\xi_0 = \xi_1 = 0$, then this problem is known as the Neumann boundary-value problem. If $\eta_0 = \eta_1 = 0$ then this is the Dirichlet boundary-value problem.

An equation of this form is usually classified as a stationary *convection–diffusion* (or *advection–diffusion*) equation. It appears in many applications related to steady heat and mass transport. For example, when this equation is applied to describe the heat conduction in a moving medium in one-dimensional space (1D), then $u = u(x)$ is the temperature, a the convection velocity, and ε the thermal diffusivity.

Let us introduce the differential operator L by the following equations:

$$\begin{cases} Lu(c) \equiv \xi_0 \, u(c) - \eta_0 \varepsilon u'(c) = \varphi, \\ Lu(x) \equiv -\varepsilon u''(x) - a(x)u'(x) + b(x)u(x) = f(x), \quad x \in (c, d), \\ Lu(d) \equiv \xi_1 \, u(d) + \eta_1 \varepsilon u'(d) = \psi. \end{cases} \tag{2.20}$$

In the operator form, this system can be written as

$$Lu(x) = F(x), \quad x \in [c, d], \tag{2.21}$$

where

$$F(x) \equiv \begin{cases} \varphi & \text{if } x = c, \\ f(x) & \text{if } x \in (c, d), \\ \psi & \text{if } x = d. \end{cases} \tag{2.22}$$

Consider a uniform grid with points $x_i \equiv c + (i - 1)h \quad (i = 1, 2, \dots, N)$, with $h = (d - c)/(N - 1)$. Let us project the problem (2.21) onto this grid, and use the difference operators (2.6)–(2.9) for approximation of derivatives $u'(x_i)$ and $u''(x_i)$ at grid points x_i:

$$u''(x_i) \approx D_+D_-u(x_i) \equiv \frac{u(x_{i+1}) - 2u(x_i) + u(x_{i-1})}{h^2},$$

$$u'(x_i) \approx \begin{cases} D_+u(x_i) \equiv (u(x_{i+1}) - u(x_i))/h, \\ D_-u(x_i) \equiv (u(x_i) - u(x_{i-1}))/h, \\ D_0 \, u(x_i) \equiv (u(x_{i+1}) - u(x_{i-1}))/2h. \end{cases} \tag{2.23}$$

This leads to the system on N algebraic equations

$$L^h u_i^h = F(x_i) \quad (i = 1, 2, \dots, N) \tag{2.24}$$

with respect to u_i^h $(i = 1, \dots, N)$, which approximate solution $u(x)$ of the problem (2.20) at the nodes of the computational grid. Such a system of *discretized equations* is usually called the *finite-difference scheme.*

For simplicity, set $\xi_0 = \xi_1 = 1$ and $\eta_0 = \eta_1 = 0$ in (2.18) and (2.19). Let us consider the following difference scheme for approximation of the differential problem (2.20):

$$\begin{cases} L^h u_1^h \equiv u_1^h = \varphi, \\ L^h u_i^h \equiv -\varepsilon\gamma_i D_+ D_- u_i^h - a(x_i) D_0 u_i^h + b(x_i) u_i^h = f(x_i), \\ \qquad\qquad\qquad\qquad i = 2, \dots, N-1, \\ L^h u_N^h \equiv u_N^h = \psi. \end{cases} \tag{2.25}$$

A family of parameters γ_i $(i = 2, \dots, N-1)$ in (2.25) plays a role of *numerical (artificial) diffusion.* As will be evident later, by a special choice of the parameters γ_i, any of the approximations (2.23) for $u'(x_i)$ in system (2.25) can be implemented. Moreover, the choice of γ_i will affect the stability of this finite-difference scheme as well as its consistency and accuracy. According to the Lax equivalence theorem, problem (2.25) must conform to the two basic principles, which are

- well-posedness, that means to possess a unique stable solution,
- consistency to the differential problem (2.20).

Let us determine the constraints on the parameters γ_i $(i = 2, \dots, N-1)$ such that these two principles are satisfied.

Stability

The system of equations (2.25) with respect to the grid function u^h can be expressed as

$$\begin{cases} L^h u_1^h \equiv B_1 u_1^h - C_1 u_2^h = F_1, \\ L^h u_i^h \equiv -A_i u_{i-1}^h + B_i u_i^h - C_i u_{i+1}^h = F_i \quad (i = 2, 3, \dots, N-1), \\ L^h u_N^h \equiv -A_N u_{N-1}^h + B_N u_N^h = F_N, \end{cases} \tag{2.26}$$

which is referred to as the *three-point problem*, and the operator L^h of this problem the *three-point operator*. Well-posedness and hence stability of this problem can be defined by using the notion of *operators of monotone type*. Operator L^h is said to be of monotone type, if the condition

$$L^h u_i^h \leqslant L^h v_i^h$$

implies

$$u_i^h \leqslant v_i^h$$

$(i = 1, 2, \dots, N)$.

Theorem: *The equation*

$$L^h u^h = f^h$$

with the operator L^h *of monotone type has a unique solution for any right-hand side* f^h.

The following theorem formulates sufficient conditions for the operator L^h to be of monotone type.

Theorem: *If coefficients in the system (2.26) such that*

$$B_1 \geqslant C_1 \geqslant 0, \quad B_N \geqslant A_N \geqslant 0, \quad B_1 \neq 0, \quad B_N \neq 0, \tag{2.27}$$

$$A_i > 0, \quad C_i > 0, \quad B_i \geqslant A_i + C_i \quad (i = 2, \dots, -1), \tag{2.28}$$

where at least one of the inequalities involving B_i $(i = 1, 2, \dots, N)$ *holds strictly, then the operator* L^h *is of monotone type.*

Therefore, if operator L^h of the finite-difference problem is of monotone type, then this ensures existence and uniqueness of the solution. Note that (2.27) and (2.28) in this theorem are conditions of diagonal dominance of the coefficient matrix of system (2.26). In order to determine conditions which L^h must satisfy to be of monotone type, rewrite the system (2.25) in the form of three-point problem:

$$\begin{cases} L^h u_1^h \equiv u_1^h = \varphi, \\ L^h u_i^h \equiv -\dfrac{\varepsilon}{h^2}\left(\gamma_i - \dfrac{a(x_i)h}{2\varepsilon}\right)u_{i-1}^h + \left(\dfrac{2\varepsilon\gamma_i}{h^2} + b(x_i)\right)u_i^h - \dfrac{\varepsilon}{h^2}\left(\gamma_i + \dfrac{a(x_i)h}{2\varepsilon}\right)u_{i+1}^h \\ \qquad = f(x_i), \\ \qquad\qquad (i = 2, \dots, N-1) \\ L^h u_N^h \equiv u_N^h = \psi. \end{cases}$$

Conditions (2.27) of the theorem are obviously satisfied. Conditions (2.28) lead to

$$\frac{\varepsilon}{h^2}(\gamma_i - 0.5\,R_i) > 0, \quad \frac{\varepsilon}{h^2}(\gamma_i + 0.5\,R_i) > 0,$$

$$\frac{2\varepsilon\gamma_i}{h^2} + b(x_i) \geqslant \frac{\varepsilon}{h^2}(\gamma_i - 0.5\,R_i) + \frac{\varepsilon}{h^2}(\gamma_i + 0.5\,R_i)$$

$(i = 2, \dots, N-1)$ or, equivalently,

$$\gamma_i - 0.5\,R_i > 0, \quad \gamma_i + 0.5\,R_i > 0, \quad b(x_i) \geqslant 0 \quad (i = 2, \dots, N-1)$$

where $R_i = a(x_i)h/\varepsilon$ denotes the local value of the grid Reynolds number (in the present context, it is equivalent to the grid Peclet number, expressing the ratio of strengths of convection and diffusion processes). Assuming that $b(x_i) \geqslant 0$ $(i = 2, \dots, N-1)$ is satisfied *a priori*, two remaining conditions require

$$\gamma_i > 0.5\,|R_i| \quad (i = 2, \ldots, N - 1). \tag{2.29}$$

This is sufficient condition, under which the difference scheme (2.25) is of monotone type. Therefore, with conditions (2.29) satisfied, numerical implementation leads to a well-posed problem, which possesses a unique solution continuously dependent on the input parameters.

Classically used finite-difference schemes are deduced by the following choice of the parameter γ_i $(i = 2, \ldots, N - 1)$.

- $\gamma_i = 1$ $(i = 2, \ldots, N - 1)$ leads to the *central difference scheme.* In this case, condition (2.29) obtains the form

  $$|R_i| < 2 \quad (i = 2, \ldots, N - 1)$$

  which leads to the constraint on the step size

  $$h < \frac{2\varepsilon}{\max\limits_{2 \leqslant i \leqslant N-1} |\, a(x_i)|} \leqslant \frac{2\varepsilon}{\|a(x)\|_\infty}.$$

  Therefore, if the diffusion coefficient ε is relatively small then this numerical scheme may violate the stability condition. This is why the central difference scheme is classified as conditionally stable.

- $\gamma_i = 1 + 0.5\,|R_i|$ $(i = 2, \ldots, N - 1)$. This method is obviously always stable, independent on the values of parameter ε and step size h. This choice of γ_i leads to the so-called *directed* (forward or backward) *difference schemes*. More specifically, if $a(x_i) > 0$ at grid point x_i then the first-order derivative $u'(x_i)$ in the difference scheme is approximated by the backward difference, otherwise, if $a(x_i) < 0$, then by the forward difference, and if $a(x_i) = 0$, then according to the central difference formula.

- $\gamma_i = 0.5\, R_i \coth(0.5\, R_i)$ $(i = 2, 3, \ldots, N - 1)$ leads to the so-called Scharfetter–Gummel's scheme or *exponential fitting scheme* [1], which is unconditionally stable.

Consistency

Let us now consider the consistency of the difference problem (2.25) and differential problem (2.20). Rewrite the system (2.25) in the form

$$L^h u_i^h = F_i^h \quad (i = 1, 2, \ldots, N)$$

where

$$F_i^h \equiv \begin{cases} \varphi & \text{if } i = 1, \\ f(x_i) & \text{if } 2 \leqslant i \leqslant N - 1, \\ \psi & \text{if } i = N. \end{cases} \tag{2.30}$$

Using definition (2.15), an estimate for consistency in the maximum norm becomes

$$\|L^h(u)^h - F^h\|_\infty = \left\| \left(L^h(u)^h - (Lu)^h\right) + \left((Lu)^h - F^h\right) \right\|_\infty$$
$$\leqslant \| L^h(u)^h - (Lu)^h \|_\infty + \| (F)^h - F^h \|_\infty .$$

Assume that no error occurs in approximating the right-hand side function $F(x)$ in (2.21), and hence the last term in this estimate vanishes. Also, assume that no error occurs at the boundary nodes $i = 1$ and $i = N$. For the remaining nodes of the grid the estimation

$$\| L^h(u)^h - (Lu)^h\|_\infty \leqslant C\left(\max_{2\leqslant i\leqslant N-1} | 1 - \gamma_i | + h^2\right) \tag{2.31}$$

can be obtained (see exercise 4). Using this estimation, accuracy of approximation of the problem (2.20) by the difference schemes considered above can be evaluated.

- For the central difference scheme, where $\gamma_i = 1$ $(i = 2, \dots, N - 1)$,

$$\| L^h(u)^h - (Lu)^h \|_\infty = O(h^2),$$

that indicates approximation of the second order in step size h.

- For the directed (forward and backward) difference scheme, where $\gamma_i = 1 + 0.5\,|R_i|$ $(i = 2, \dots, N - 1)$, from

$$\max_{2\leqslant i\leqslant N-1} | 1 - \gamma_i | = \max_{2\leqslant i\leqslant N-1} | R_i | \leqslant \frac{\|a(x)\|_\infty}{2\varepsilon} h$$

it follows that

$$\| L^h(u)^h - (Lu)^h\|_\infty \leqslant C\left(\frac{\|a(x)\|_\infty}{2\varepsilon} h + h^2\right) = O(h),$$

that indicates the first-order accuracy.

- For the exponential scheme, where $\gamma_i = 0.5\, R_i \coth(0.5\, R_i)$ $(i = 2, 3, \dots, N - 1)$, it can be shown (see exercise 4) that the approximation accuracy is of second order in h.

2.1.3 Numerical solution of a system with three-diagonal matrix: Thomas (TDMA) algorithm

Solution of the difference problem (2.25), corresponding to the differential problem (2.20), as can be seen from previous section, leads to the following system of linear algebraic equations:

$$\begin{cases} B_1 u_1 - C_1 u_2 = F_1, & \\ -A_i u_{i-1} + B_i u_i - C_i u_{i+1} = F_i & (i = 2, 3, \dots, N - 1), \\ -A_N u_{N-1} + B_N u_N = F_N. & \end{cases} \tag{2.32}$$

Solution of this system $\{u_i\}$ $(i = 1, \dots, N)$ approximates values of the exact solution $u(x)$ of the corresponding differential problem at nodes $\{x_i\}$ $(i = 1, \dots, N)$ of the computational grid. The first and the last equations in system (2.32) approximate boundary conditions, while the equations from the second to the next to last $(i = 2, \dots, N - 1)$ approximate values of $u(x)$ at the internal nodes x_i $(i = 2, \dots, N - 1)$ of the grid.

Non-zero entries of the coefficient matrix of system (2.32) locate along the three diagonals. Indeed,

- entries B_i are along the main diagonal of the matrix,
- entries C_i are along the diagonal just above the main diagonal, and
- entries A_i are along the diagonal just below the main diagonal,

$$\begin{bmatrix} B_1 & -C_1 & 0 & \cdots & \cdots & \cdots & \cdots & \cdots & 0 & 0 \\ -A_2 & B_2 & -C_2 & \cdots & \cdots & \cdots & \cdots & \cdots & 0 & 0 \\ \cdots & \cdots & \cdots & \cdots & \cdots & \cdots & \cdots & \cdots & \cdots & \cdots \\ 0 & 0 & 0 & \cdots & -A_i & B_i & -C_i & \cdots & 0 & 0 \\ \cdots & \cdots & \cdots & \cdots & \cdots & \cdots & \cdots & \cdots & \cdots & \cdots \\ 0 & 0 & 0 & \cdots & \cdots & \cdots & \cdots & -A_{N-1} & B_{N-1} & -C_{N-1} \\ 0 & 0 & 0 & \cdots & \cdots & \cdots & \cdots & 0 & -A_N & B_N \end{bmatrix} \cdot \begin{bmatrix} u_1 \\ u_2 \\ \cdots \\ u_i \\ \cdots \\ u_{N-1} \\ u_N \end{bmatrix} = \begin{bmatrix} F_1 \\ F_2 \\ \cdots \\ F_i \\ \cdots \\ F_{N-1} \\ F_N \end{bmatrix}$$

Adaptation of the Gauss elimination method to this system with a tridiagonal coefficient matrix is known as Thomas algorithm (also abbreviated to TDMA). From the first equation of (2.32),

$$u_1 = \frac{C_1}{B_1}u_2 + \frac{F_1}{B_1} \equiv \alpha_1 u_2 + \beta_1. \tag{2.33}$$

Eliminating u_1 from the second equation of the system, we have

$$u_2 = \frac{C_2}{B_2 - \alpha_1 A_2}u_3 + \frac{F_2 + \beta_1 A_2}{B_2 - \alpha_1 A_2} \equiv \alpha_2 u_3 + \beta_2.$$

At the $(i - 1)$ step of the elimination process,

$$u_{i-1} = \alpha_{i-1}u_i + \beta_{i-1}. \tag{2.34}$$

Using this to eliminate u_{i-1} from the ith equation of the system, we have

$$u_i = \frac{C_i}{B_i - \alpha_{i-1}A_i}u_{i+1} + \frac{F_i + \beta_{i-1}A_i}{B_i - \alpha_{i-1}A_i} \equiv \alpha_i u_{i+1} + \beta_i.$$

Eventually, from

$$u_{N-1} = \alpha_{N-1}u_N + \beta_{N-1}$$

and the last equation of system (2.32) it follows

$$u_N = \frac{F_N + \beta_{N-1}A_N}{B_N - \alpha_{N-1}A_N}. \tag{2.35}$$

Equations (2.33), (2.34), and (2.35) form the Thomas algorithm, which consists of two basic stages:

I. Elimination stage (calculation of coefficients α_i, β_i):

$$\left[\begin{array}{l} \alpha_1 = C_1/B_1, \quad \beta_1 = F_1/B_1 \\ \alpha_i = C_i/(B_i - \alpha_{i-1}A_i), \quad \beta_i = (F_i + \beta_{i-1}A_i)(B_i - \alpha_{i-1}A_i) \quad (i = 2, \dots, N-1). \end{array}\right.$$

II. Back substitution stage (obtaining the solution u_i):

$$\left[\begin{array}{lcl} u_N & = & (F_N + \beta_{N-1}A_N)/(B_N - \alpha_{N-1}A_N), \\ u_i & = & \alpha_i u_{i+1} + \beta_i, \quad (i = N-1, N-2, \dots, 1). \end{array}\right.$$

MATLAB/Octave [2] implementation of this algorithm is presented below, in the form of function named `TDMA`. Given that one-dimensional arrays `A`, `B`, `C`, and `F`, each of size `(1:N)`, define corresponding entries of the three-diagonal matrix, then solution `u` is obtained by calling this function by the command `u = TDMA(A,B,C,F)`.

```
function u = TDMA(a, b, c, f)
n = length(f);
%1st stage of the algorithm
alf(1) = c(1)/b(1);
bet(1) = f(1)/b(1);
for i = 2: n - 1
  dummy = b(i) - alf(i - 1)*a(i);
  alf(i) = c(i)/dummy;
  bet(i) = (f(i) + bet(i - 1)*a(i))/dummy;
end
%2nd stage of the algorithm
u(n) = (f(n) + bet(n - 1)*a(n))/(b(n) - alf(n - 1)*a(n));
for i = n - 1: -1: 1
  u(i) = alf(i)*u(i + 1) + bet(i);
end
```

Exercise 4 (Finite-difference schemes)

1. Derive the estimate (2.31).
2. Show that the exponential fitting scheme leads to the second order convergence $\|(u)^h - u^h\|_\infty \leqslant Ch^2$.
3. Derive approximations for the following boundary conditions in the form consistent with the first and last equations in system (2.26). Assume that

$x = a$ and $x = b$ stand for the left and right end points of the computational domain, α and β, m and n are real parameters.

a) $u(a) = \alpha, \quad u(b) = \beta$;
b) $u'(a) = \alpha, \quad u'(b) = \beta$;
c) $u'(a) - mu(a) = \alpha, \quad u'(b) + nu(a) = \beta$.

4. Find the numerical solution of the boundary-value problem

$$-\varepsilon u'' - xu' + u = (1 + \varepsilon\pi^2)\cos(\pi x) + \pi\, x\sin(\pi x),$$
$$u(-1) = -1, \quad u(1) = 1,$$

by using

a) central differences scheme;
b) directed differences scheme;
c) exponential scheme.

Use the uniform grid $x_i = a + (i - 1)h$ ($i = 1, 2, \dots, N$), with step size $h = (b - a)/(N - 1)$, where $x = a$ and $x = b$ are the end points of the computational domain. In each of the cases (a)–(c), examine

- the effect of the grid refinement (set N= 10, 20, 40, and 80 nodes at $\varepsilon = 0.01$), and
- the effect of decrease of the parameter ε (set $\varepsilon = 10^0, 10^{-1}, 10^{-2}$ and 10^{-3} at $N = 40$) on the accuracy of the numerical solution. The computational errors can be estimated by using the *RMS* (round mean square) and the *maximum* norms:

$$\|e\|_2 = \left[\frac{1}{N}\sum_{i=1}^{N}(u(x_i) - u_i)^2\right]^{1/2} \quad \text{and} \quad \|e\|_\infty = \max_{1\leqslant i\leqslant N} | u(x_i) - u_i |,$$

where u_i approximate exact solution $u(x_i)$ at nodes x_i ($i = 1, 2, \dots, N$) of the grid.

[Analytic solution: $u(x) = \cos(\pi x) + x + \frac{x\,\mathrm{erf}(x/\sqrt{2\varepsilon}) + \sqrt{2\varepsilon/\pi}\exp(-x^2/2\varepsilon)}{\mathrm{erf}(1/\sqrt{2\varepsilon}) + \sqrt{2\varepsilon/\pi}\exp(-1/2\varepsilon)}$]

2.1.4 Numerical methods of solution of non-linear (quasi-linear) convection–diffusion equation

Consider the second order quasi-linear differential equation of the following form (compare it to equation (2.17) considered in section 2.1.2):

$$-\varepsilon u''(x) - a(x)u'(x) + g(u(x)) = f(x). \tag{2.36}$$

Numerical solution of non-linear problems usually consists of two stages:

- linearization of the problem, and
- solution of the linearized problem using suitable method.

These two stages are repeated within the iteration loop, until convergence with desired accuracy is achieved. Consider the following iterative methods.

Method 1: First, consider the approach based on the *Newton's iteration method.* Let $u^{(n)}(x)$ be a solution of the differential equation (2.36) obtained from the nth iteration. Using Taylor's expansion, non-linear term in the differential equation can be written

$$\begin{aligned} g(u^{(n+1)}) &= g(u^{(n)} + \Delta u) \\ &= g(u^{(n)}) + \Delta u \cdot g_u(u^{(n)}) + \text{HOT}, \end{aligned} \tag{2.37}$$

where

$$\Delta u = u^{(n+1)} - u^{(n)}.$$

Introducing (2.37) for $g(u(x))$ in equation (2.36), it reduces to a linear equation with respect to the $(n + 1)$ iterative approximation of the solution $u^{(n+1)}$:

$$-\varepsilon\frac{d^2u^{(n+1)}}{dx^2} - a(x)\frac{du^{(n+1)}}{dx} + g_u(u^{(n)})u^{(n+1)} = f(x) - g(u^{(n)}) + g_u(u^{(n)})u^{(n)}, \tag{2.38}$$

which can be solved by any of the methods from section 2.1.2, designed for linear differential equations, and by applying the TDMA.

Method 2: Algorithm based on the Newton's iteration method can be applied to the difference scheme rather than the differential equation. Consider the three-point problem of the form (2.26) and define the operator $\mathcal{G}$ as follows:

$$\begin{cases} \mathcal{G}_1(u^h) \equiv B_1u_1^h - C_1u_2^h - F_1 = 0, \\ \mathcal{G}_i(u^h) \equiv -A_iu_{i-1}^h + B_iu_i^h - C_iu_{i+1}^h - F_i = 0 \quad (i = 2, 3, \dots, N-1) \\ \mathcal{G}_N(u^h) \equiv -A_Nu_{N-1}^h + B_Nu_N^h - F_N = 0, \end{cases} \tag{2.39}$$

or, equivalently,

$$\mathcal{G}(u^h) = 0 \tag{2.40}$$

in the operator form. Here, u^h is the grid function approximating solution $u(x)$ of the equation (2.36) at points of the computational grid. Assume that the $(n + 1)$st iterative solution $u^{h,n+1}$ of the problem (2.39) is sufficiently close to the exact solution u^h of this problem,

$$u^h \approx u^{h,\,n+1} = u^{h,\,n} + \Delta u^{h,\,n}.$$

Then, keeping in the expansion a linear part only, we have

$$\mathcal{G}(u^{h,\,n} + \Delta u^{h,\,n}) = \mathcal{G}(u^{h,\,n}) + \mathcal{G}'(u^{h,\,n})\Delta u^{h,\,n} = 0,$$

from which

$$\mathcal{G}'(u^{h,\,n})\Delta u^{h,\,n} = -\mathcal{G}(u^{h,\,n}). \tag{2.41}$$

This is a linear system of N equations for grid function $\Delta u^{h,\,n} = u^{h,\,n+1} - u^{h,\,n}$, with three-diagonal coefficient matrix (more precisely, the Jacobian)

$$\mathcal{G}'(u^{h,\,n}) = \left[\frac{\partial \mathcal{G}_i}{\partial u_j^h}\right] \quad (i, j = 1, \, \ldots \, , N).$$

Therefore, the system (2.41) can be solved by using TDMA described in section 2.1.3, within the iteration loop, until desired accuracy $\left\|\Delta u^{h,\,n}\right\|_p < \delta$ is achieved with sufficiently small δ.

Method 3: Finally, consider the *simple-iteration method* based on the solution of the differential equation

$$-\varepsilon\frac{d^2u^{(n+1)}}{dx^2} - a(x)\frac{du^{(n+1)}}{dx} + g(u^{(n)}) = f(x), \tag{2.42}$$

given the solution $u^{(n)}(x)$ from the previous iteration. Again, methods from section 2.1.2 and TDMA can be applied to compute the solution $u^{(n+1)}(x)$.

Exercise 5 (Quasi-linear equation)

1. Find the numerical solution of the boundary-value problem

$$-\varepsilon u'' + u + u^2 = \exp(-2x/\sqrt{\varepsilon}), \quad u(0) = 1, \quad u(1) = \exp(-1/\sqrt{\varepsilon}),$$

by the iterative **Methods 1–3**. Use uniform grid $x_i = a + (i - 1)\,h$ $(i = 1, 2, \, \ldots \, , N)$, with step size $h = (b - a)/(N - 1)$, where $x = a$ and $x = b$ are the end points of the computational domain. Examine the rates of convergence of the iterative methods

- varying N= 10, 20, 40, and 80 at $\varepsilon = 0.01$, and
- varying $\varepsilon = 10^0, 10^{-1}, 10^{-2}$, and 10^{-3} at $N = 40$.

The computational errors in the *RMS* and the *maximum* norms at nth iteration can be evaluated by using formulas

$$\|e^{(n)}\|_2 = \left[\frac{1}{N}\sum_{i=1}^{N}\left(u(x_i) - u_i^{(n)}\right)^2\right]^{1/2} \quad \text{and} \quad \|e^{(n)}\|_\infty = \max_{1\leqslant i\leqslant N} | \, u(x_i) - u_i^{(n)} \, |,$$

where $u_i^{(n)}$ approximate exact solution $u(x_i)$ at nodes x_i $(i = 1, 2, \, \ldots \, , N)$ of the grid. Note that convergence of the method can be estimated also from $\left\|u^{(n+1)} - u^{(n)}\right\|_p$ by using two successive iterative solutions $u_i^{(n)}$ and $u_i^{(n+1)}$ $(i = 1, 2, \, \ldots \, , N)$. [Analytic solution: $u(x) = \exp(-x/\sqrt{\varepsilon})$]

2.2 Finite volume method (FVM) for convection–diffusion equation

In this section, we return to the one-dimensional drift–diffusion equation (see equation (2.3))

$$\frac{\partial n}{\partial t} + \frac{\partial}{\partial x}\left(vn - D\frac{\partial n}{\partial x}\right) = S. \tag{2.43}$$

As we have seen from chapter 1, this equation, with $v = \text{sign}(q)\mu E$ defining the drift velocity and D the diffusion coefficient, can be used to describe the time evolution of the plasma species density, where the analysis is restricted to one space dimension. In more general terms, this equation can be classified as a *convection–diffusion* (or *transport*) equation. In the following, the FVM for this equation is described systematically. We start from the simplest special case of this equation, namely,

- *the steady diffusion equation,*

and then sequentially increase the complexity, considering one by one

- *the steady convection and diffusion equation,*
- *the time-dependent diffusion equation,* and eventually
- *the time-dependent convection and diffusion equation.*

2.2.1 Steady diffusion equation

Let us consider first the steady diffusion equation, which is obtained by neglecting the convection flux from equation (2.43) and assuming that the process is steady,

$$\frac{d}{dx}\left(D\frac{dn}{dx}\right) + S = 0. \tag{2.44}$$

In terms of the flux density $\Gamma = -D\, dn/dx$, this equation obtains the form

$$\frac{d\Gamma}{dx} = S. \tag{2.45}$$

Integrating this equation over the control volume (see figure 2.2),

$$\int_{x_w}^{x_e}\left(\frac{d\Gamma}{dx}\right)dx = \int_{x_w}^{x_e} Sdx.$$

we have

$$\Gamma_e - \Gamma_w = S_P\Delta x,$$

and finally

$$\left(D\frac{dn}{dx}\right)_e - \left(D\frac{dn}{dx}\right)_w + S_P\Delta x = 0.$$

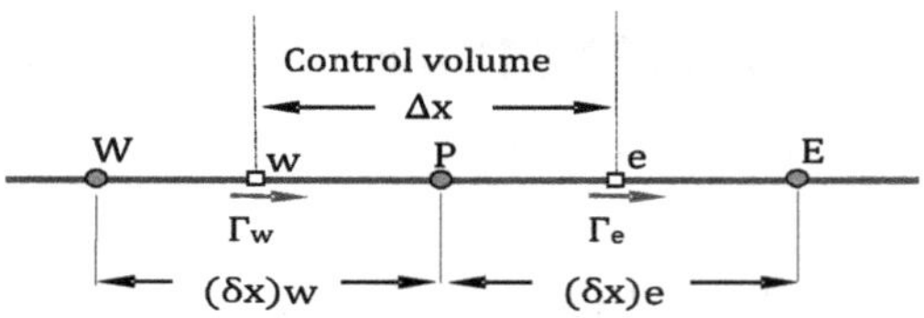

Figure 2.2. Schematic diagram of the one-dimensional control volume.

Here, Γ_e and Γ_w denote the flux densities evaluated at *east* and *west* faces x_e and x_w of the control volume, S_P is the value of the source term S at grid point P.

Let us approximate the diffusion flux $D\,dn/dx$ at faces x_e and x_w of the control volume by the central difference,

$$D_e\frac{n_E - n_P}{(\delta x)_e} - D_w\frac{n_P - n_W}{(\delta x)_w} + S_P\Delta x = 0.$$

Denoting the diffusion conductance at faces of the control volume by

$$G_e = \frac{D_e}{(\delta x)_e} \text{ and } G_w = \frac{D_w}{(\delta x)_w},$$

the difference equation corresponding to equation (2.44) obtains the form of *the three-point problem*

$$a_P n_P = a_E n_E + a_W n_W + b, \tag{2.46}$$

where

$$a_E = G_e, \quad a_W = G_w, \quad a_P = a_E + a_W, \quad b = S_P\Delta x.$$

Let us number the grid points (and also values of the density n at these points) with index i ($i = 1, \ldots, N$), such that indices $i = 1$ and $i = N$ correspond to the left and right ends of the domain, respectively. Then, supplying nodes x_W, x_P, and x_E (see figure 2.2) with indices $i - 1$, i, and $i + 1$, respectively, and also introducing notations A_i, B_i, and C_i for coefficients of n at these points, the three-point problem (2.46) obtains its equivalent form

$$-A_i n_{i-1} + B_i n_i - C_i n_{i+1} = F_i,$$

where the index i varies over the range ($i = 2, 3, \ldots, N - 1$) covering the internal nodes of the grid. This set of ($N - 2$) equations is completed with two additional equations approximating boundary conditions for diffusion equation (2.44) at the ends of the domain. Thus, the diffusion problem reduces to the solution of system of N linear algebraic equations with respect to N values n_i ($i = 1, 2, \ldots, N$), with the three-diagonal matrix coefficient, suitable for TDMA (see section 2.1.3).

Exercise 6 (Steady diffusion equation)

1. Find the numerical solution of the boundary-value problem

   $$-\varepsilon u'' + u = (\varepsilon\pi^2 + 1)\cos(\pi x), \quad u(-1) = 0, \quad u(1) = -1.$$

   [Analytic solution: $u(x) = \cos(\pi x) + \sinh((1 - x)/\sqrt{\varepsilon})/\sinh(2/\sqrt{\varepsilon})$]
2. Find the numerical solution of the steady diffusion problem (2.44), where $D = 0.02$ m^2/(s), $S = a\exp[-(x - b)^2/c^2]$, and the length of the computational domain $d = 1$ cm. Apply the boundary conditions $n(0) = 0$ and $n'(d) = 0$. Use $a = 4.6 \times 10^{23}$ m^{-3} s^{-1}, $b = 0.8$ cm, and $c = 0.1$ cm. Numerical solution $n(x)$ corresponding to these parameters is shown in figure 2.3(a). This figure also contains the source term $S(x)$.

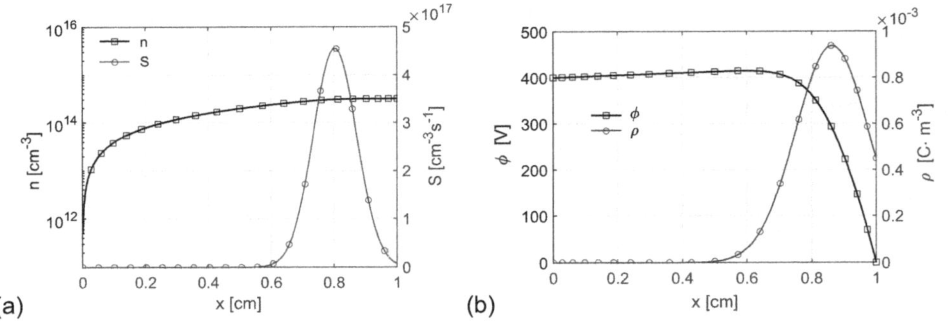

Figure 2.3. (a) solution of problem 2 and (b) problem 3 from exercise 6.

3. Find the numerical solution of Poisson equation $-\varepsilon_0\, d^2\varphi/dx^2 = \rho(x)$, on the interval $0 < x < d$, where the charge density $\rho = a\exp[-(x-b)^2/c^2]$. Apply the boundary conditions $\varphi(0) = V_0$, $\varphi(d) = 0$. Let $d = 1$ cm and $V_0 = 400$ V. Use $a = 9.4 \times 10^{-4}$ cm^{-3}, $b = 0.86$ cm, and $c = 0.16$ cm. Numerical solution $\varphi(x)$ corresponding to these parameters and the charge density profile $\rho(x)$ are shown in figure 2.3(b).

2.2.2 Steady convection and diffusion equation

The convection and diffusion equation (2.43) in steady state obtains the form

$$\frac{d}{dx}\left(vn - D\frac{dn}{dx}\right) = S, \tag{2.47}$$

or

$$\frac{d\Gamma}{dx} = S \tag{2.48}$$

in terms of the total flux density $\Gamma = nv - D\frac{\partial n}{\partial x}$. The *staggered grid* (see figure 2.2), such that the scalar variable (the density n) is assigned to the cell centers of the control volumes (the points W, E, and P in figure 2.2), whereas the vector variable (the convection velocity v) to the cell faces (the points w and e), is used. Integrating equation (2.48) over the control volume ($x_w \leqslant x \leqslant x_e$) (see figure 2.2),

$$\int_{x_w}^{x_e}\left(\frac{d\Gamma}{dx}\right)dx = \int_{x_w}^{x_e} Sdx.$$

we obtain

$$\Gamma_e - \Gamma_w = S_P\Delta x, \tag{2.49}$$

where the subscripts e and w denotes values at the faces x_e and x_w of the control volume. Rewrite equation (2.49) in the form

$$\left(vn - D\frac{dn}{dx}\right)_e - \left(vn - D\frac{dn}{dx}\right)_w = S_P \Delta x,$$

and, equivalently,

$$(vn)_e - (vn)_w - \left[\left(D\frac{dn}{dx}\right)_e - \left(D\frac{dn}{dx}\right)_w\right] = S_P \Delta x. \tag{2.50}$$

A reasonable way to approximate the diffusion flux $-Ddn/dx$ at faces of the control volume x_e and x_w is by the central differences:

$$\left(D\frac{dn}{dx}\right)_e = D_e \frac{n_E - n_P}{(\delta x)_e} \quad \text{and} \quad \left(D\frac{dn}{dx}\right)_w = D_w \frac{n_P - n_W}{(\delta x)_w}.$$

In fact, this approach is ideally suited to the case of piecewise-linear shape of the density n on the given grid. The actual form of the numerical scheme is determined by the approximation of the drift (convection) flux vn at points x_e and x_w. Recall that density n is defined at the nodes of the grid (points W, P, and E in figure 2.2), while the convection velocity v at the cell faces x_w and x_e. Consider the following three methods.

- **Central differences scheme:** Consider a piecewise-linear profile for the density n spanned by the numerical grid. In this situation, if the volume faces x_e and x_w locate midway between the grid points, densities at volume faces are the arithmetic means of their values at the grid points, and equation (2.50) obtains the form

$$v_e \frac{n_E + n_P}{2} - v_w \frac{n_P + n_W}{2} - \left[D_e \frac{n_E - n_P}{(\delta x)_e} - D_w \frac{n_P - n_W}{(\delta x)_w}\right] = S_P \Delta x. \tag{2.51}$$

With notations

$$a_E = G_e - \frac{v_e}{2}, \quad a_W = G_w + \frac{v_w}{2}, \quad a_P = G_e + \frac{v_e}{2} + G_w - \frac{v_w}{2} = a_E + a_W + (v_e - v_w),$$

$$b = S_P \Delta x,$$

equation (2.51) is written in the form of a three-point problem

$$a_P n_P = a_E n_E + a_W n_W + b.$$

- **Upwind method:** According to this method, value of density n at an interface of the control volume is specified equal to its value at the grid point on the upwind side of the face:

$$n_e = \begin{cases} n_P & \text{if} \quad v_e > 0, \\ n_E & \text{if} \quad v_e < 0, \end{cases}$$

and similar for n_w:

$$n_w = \begin{cases} n_W & \text{if} \quad v_w > 0, \\ n_P & \text{if} \quad v_w < 0. \end{cases}$$

Let us denote $[a, b] = \max(a, b)$, the greater of a and b. Then, the convection fluxes vn at points x_e and x_w in equation (2.50) are expressed by

$$(vn)_e = n_P[v_e, 0] - n_E[-v_e, 0] \quad \text{and} \quad (vn)_w = n_W[v_w, 0] - n_P[-v_w, 0].$$

Therefore, the coefficients in the three-point problem

$$a_P n_P = a_E n_E + a_W n_W + b$$

obtain the form

$$a_E = G_e + [-v_e, 0], \qquad a_W = G_w + [v_w, 0],$$

$$a_P = G_e + [v_e, 0] + G_w + [-v_w, 0] = a_E + a_W + (v_e - v_w), \qquad b = S_P \Delta x.$$

- **Exponential scheme:** Let us define by

$$P_e = \frac{v_e}{G_e} \quad \text{and} \quad P_w = \frac{v_w}{G_w}$$

the grid Peclet numbers [1], measuring the ratios of strengths of the convection and diffusion fluxes at faces x_e and x_w of the control volume. Then, the coefficients in the three-point problem

$$a_P n_P = a_E n_E + a_W n_W + b \tag{2.52}$$

are defined (for details, see [1]) as follows

$$a_E = \frac{G_e P_e}{\exp(P_e) - 1}, \qquad a_W = \frac{G_w P_w \exp(P_w)}{\exp(P_w) - 1}, \qquad a_P = a_E + a_W + (v_e - v_w),$$

$$b = S_P \Delta x.$$

Therefore, by any of these three methods, numerical implementation of the convection–diffusion equation (2.47) reduces to the three-point problem of the form (2.32), which can be solved by using the TDMA algorithm.

Remark. *Numerical implementation can include all of these three methods, namely, central differences scheme, upwind method, and exponential scheme, combined within the same framework [1]. In this situation,*

$$a_E = G_e\, A(|P_e|) + [-v_e, 0], \qquad a_W = G_w\, A(|P_w|) + [v_w, 0],$$

a_P *and* b *are defined as described above,*

$$a_P = a_E + a_W + (v_e - v_w), \qquad b = S_P \Delta x,$$

and function $A(|P|)$ *is specified as in table* 2.1.

Table 2.1. Definition of function $A(|P|)$ for different numerical schemes [1].

No	Method	$A(\|P\|)$
1	Central differences scheme	$1 - 0.5\,\|P\|$
2	Upwind method	1
3	Exponential scheme	$\|P\|/[\exp(\|P\|) - 1]$

Exercise 7 (Steady convection and diffusion equation)

1. Show that the generalized formulation leads to (i) central differences scheme, (ii) upwind method, and (iii) exponential scheme, subject to definitions of function $A(|P|)$ according to table 2.1.
2. Use definitions in section 2.1.1 and corresponding theorems in 2.1.2 to explore the *consistency*, *stability*, and *convergence* of the
   a) central differences scheme;
   b) upwind method;
   c) exponential scheme.
3. Find the numerical solution of the boundary-value problem

$$-(xu + \varepsilon u')' + u = 0, \quad u(-1) = 0, \quad u(1) = 2,$$

by using
   a) central differences scheme,
   b) upwind method,
   c) exponential scheme.

Use staggered grid with points $x_i = a + (i - 1)\,h$ $(i = 1, 2, \dots, N)$, where step size $h = (b - a)/(N - 1)$, and $x = a$ and $x = b$ are the end points of the computational domain. In each of the cases (a)–(c), examine

- the effect of the grid refinement (set $N=$ 10, 20, 40, and 80 nodes at $\varepsilon = 0.01$) on the accuracy of the numerical solution, and
- the effect of decrease of the parameter ε (set $\varepsilon = 10^0, 10^{-1}, 10^{-2}$ and 10^{-3} at $N = 40$) on the accuracy of the numerical solution.

The computational error in the *RMS* and the *maximum* norms can be evaluated by using the following formulas:

$$\|e\|_2 = \left[\frac{1}{N}\sum_{i=1}^{N}(u(x_i) - u_i)^2\right]^{1/2} \quad \text{and} \quad \|e\|_\infty = \max_{1\leqslant i\leqslant N} |\, u(x_i) - u_i \,|,$$

where u_i approximates exact solution $u(x_i)$ at nodes x_i $(i = 1, 2, \dots, N)$ of the grid.

[Analytic solution: $u(x) = 1 + \mathrm{erf}(x/\sqrt{2\varepsilon})/\,\mathrm{erf}(1/\sqrt{2\varepsilon})$]

4. Find the numerical solution of the steady form of the drift–diffusion problem (2.47) for positive ions, by using
   (a) central differences scheme,
   (b) upwind method,
   (c) exponential scheme.

   Define the mobility and diffusion coefficients by $\mu p = 0.80$ m^2 Torr/(sV) and $D = \mu k_B T/e$. Let the source term $S = a \exp[-(x - b)^2/c^2]$ with $a = 4.6 \times 10^{23}$ m^{-3} s^{-1}, $b = 0.8$ cm, and $c = 0.1$ cm. The electric field strength $E = 3.4 \exp(675\, x)$ V/m. Set the length of the computational domain $d = 1$ cm, the pressure $p = 1$ Torr, and the temperature $T = 290$ K. Apply the boundary conditions $n(0) = 0$ and $n'(d) = 0$. The numerical solution $n(x)$ is shown in figure 2.4. This figure also contains profiles of the source term $S(x)$ and the electric field $E(x)$.

2.2.3 Time-dependent diffusion equation

The time-dependent diffusion equation in one dimension has the form

$$\frac{\partial n}{\partial t} - \frac{\partial}{\partial x}\left(D\frac{\partial n}{\partial x}\right) = S, \tag{2.53}$$

or, equivalently,

$$\frac{\partial n}{\partial t} + \frac{\partial \Gamma}{\partial x} = S, \tag{2.54}$$

where $\Gamma = -D\frac{\partial n}{\partial x}$ is the flux density. Integrating this equation over the control volume $x_w \leqslant x \leqslant x_e$ (see figure 2.2), and also over the time interval from some t to $t + \Delta t$, we obtain

$$\int_t^{t+\Delta t}\int_{x_w}^{x_e}\left(\frac{\partial n}{\partial t}\right)dtdx + \int_t^{t+\Delta t}\int_{x_w}^{x_e}\left(\frac{\partial \Gamma}{\partial x}\right)dtdx = \int_t^{t+\Delta t}\int_{x_w}^{x_e} S\, dtdx. \tag{2.55}$$

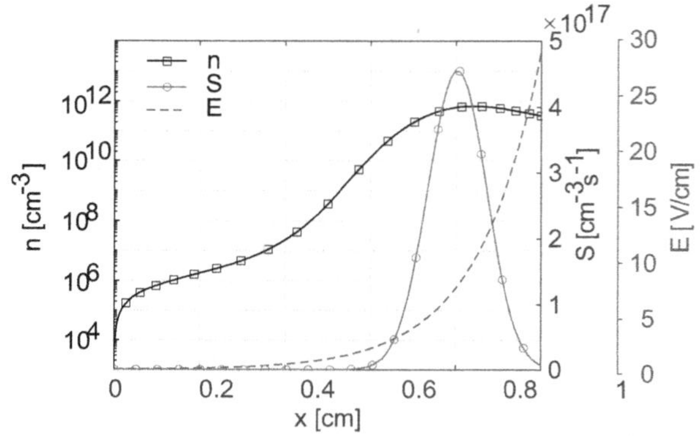

Figure 2.4. Solution of problem 4 from exercise 7.

The right-hand side of this equation is approximated by the value of the source term S at grid point P times the volume size Δx and the time step Δt,

$$\int_t^{t+\Delta t} \int_{x_w}^{x_e} S\, dtdx = S_P \Delta t \Delta x.$$

The first term in the left-hand side is approximated as

$$\int_{x_w}^{x_e} \left(\int_t^{t+\Delta t} \frac{\partial n}{\partial t} dt \right) dx = \int_{x_w}^{x_e} (n|_{t+\Delta t} - n|_t) dx = (n_P^1 - n_P^0)\Delta x,$$

where the subscript 0 indicates value of the density n at the current time level and superscripts 1 at the next time level. In fact, the actual form of the numerical scheme depends on the approximation of fluxes over the time interval from t to $t + \Delta t$ in the second term of equation (2.55). Consider the following methods.

- **Fully implicit scheme:** Let us approximate the fluxes by their values from the new time level $t + \Delta t$:

$$\int_t^{t+\Delta t} \left(\int_{x_w}^{x_e} \frac{\partial \Gamma}{\partial x} dx \right) dt = \int_t^{t+\Delta t} (\Gamma|_{x_e} - \Gamma|_{x_w}) dt = (\Gamma_e - \Gamma_w)^1 \Delta t,$$

where $\Gamma_e = \Gamma|_{x_e}$ and $\Gamma_w = \Gamma|_{x_w}$. Equation (2.55) obtains the form

$$(n_P^1 - n_P^0)\Delta x + (\Gamma_e - \Gamma_w)^1 \Delta t = S_P \Delta x \Delta t,$$

and

$$\frac{n_P^1 - n_P^0}{\Delta t}\Delta x - D_e \frac{n_E^1 - n_P^1}{(\delta x)_e} + D_w \frac{n_P^1 - n_W^1}{(\delta x)_w} = S_P \Delta x. \tag{2.56}$$

Denoting the diffusion conductances $G_e = D_e/(\delta x)_e$ and $G_w = D_w/(\delta x)_w$, and omitting superscripts 1, equation (2.56) obtains the form

$$\frac{n_P - n_P^0}{\Delta t}\Delta x - G_e(n_E - n_P) + G_w(n_P - n_W) = S_P \Delta x, \tag{2.57}$$

and, eventually,

$$a_P n_P = a_E n_E + a_W n_W + b, \tag{2.58}$$

which is a three-point problem, where

$$a_E = G_e,\ a_W = G_w,\ a_P = a_E + a_W + \frac{\Delta x}{\Delta t},\ b = S_P \Delta x + n_P^0 \frac{\Delta x}{\Delta t}. \tag{2.59}$$

This numerical scheme is called *fully implicit* according to the time progression. Within this method, numerical solution of the time-dependent diffusion equation at each time level reduces to solution of the three-point problem of the form (2.32). Therefore, time evolution of the density profile n can be obtained by successive implementation of the TDMA, starting from the initial density profile.

- **Explicit scheme:** Marching in time can be implemented explicitly that makes it unnecessary to resolve the system of equations at each time level and hence to use TDMA. In this case, fluxes Γ_e and Γ_w in equation (2.55) are taken from the current time level, so that this equation reduces to

$$\frac{n_P - n_P^0}{\Delta t}\Delta x + (\Gamma_e - \Gamma_w)^0 = S_P \Delta x, \tag{2.60}$$

from which

$$\frac{n_P - n_P^0}{\Delta t}\Delta x - D_e\frac{n_E^0 - n_P^0}{(\delta x)_e} + D_w\frac{n_P^0 - n_W^0}{(\delta x)_w} = S_P \Delta x, \tag{2.61}$$

and, eventually,

$$a_P n_P = b, \tag{2.62}$$

where $b = a_E n_E^0 + a_W n_W^0 + (a_P - a_E - a_W)n_P^0 + S_P\Delta x$, a_E and a_W are defined as in (2.59), and $a_P = \Delta x/\Delta t$.

- **Weighted scheme:** Fully implicit and explicit methods can be combined within the framework of the weighted method,

$$\frac{n_P - n_P^0}{\Delta t}\Delta x + \theta(\Gamma_e - \Gamma_w) + (1 - \theta)(\Gamma_e - \Gamma_w)^0 = S_P \Delta x, \tag{2.63}$$

where the weighting factor θ varies between $0 \leqslant \theta \leqslant 1$. This equation is obtained from (2.55) by weighting values of fluxes Γ_e and Γ_w at current and new time levels. Note that with $\theta = 0$ this scheme reduces to the explicit method, while with $\theta = 1$ to the fully implicit method. Then, from

$$\begin{aligned}\frac{n_P - n_P^0}{\Delta t}\Delta x = {} & \theta\left[D_e\frac{n_E - n_P}{(\delta x)_e} - D_w\frac{n_P - n_W}{(\delta x)_w}\right] \\ & + (1 - \theta)\left[D_e\frac{n_E^0 - n_P^0}{(\delta x)_e} - D_w\frac{n_P^0 - n_W^0}{(\delta x)_w}\right] + S_P\Delta x\end{aligned}$$

we have

$$\begin{aligned}a_P n_P = {} & a_E[\theta\, n_E + (1 - \theta)\, n_E^0] + a_W[\theta\, n_W + (1 - \theta)\, n_W^0] \\ & + [a_P^0 - (1 - \theta)a_E - (1 - \theta)\, a_W]n_P^0 + S_P\Delta x,\end{aligned}$$

from which the three-point problem can be obtained in the form

$$\begin{aligned}a_P n_P = {} & \theta\, a_E n_E + \theta\, a_W\, n_W \\ & + \Big\{(1 - \theta)a_E\, n_E^0 + (1 - \theta)a_W\, n_W^0 + [a_P^0 - (1 - \theta)a_E - (1 - \theta)\, a_W]n_P^0 \\ & + S_P\Delta x\Big\},\end{aligned}$$

where a_E and a_W are defined as in equation (2.59), $a_P^0 = \Delta x/\Delta t$, and $a_P = \theta\,(a_E + a_E) + a_P^0$.

Remark.

1. *With $\theta = 0.5$ the weighted scheme is known as the Crank–Nicolson method.*
2. *The concepts of consistency, stability, and convergence introduced in section 2.1.1 can be extended in much the same way to the case of evolutionary problems. For details, see [3, 4].*

Exercise 8 (Time-dependent diffusion equation)

1. Explore the consistency, stability, and convergence of the explicit, implicit, and weighted schemes for the diffusion equation (2.53). Consider the homogeneous equation ($S \equiv 0$) with constant diffusion coefficient D. (The stability can be studied by the Von Neumann method, see e.g., [3].) Show that
   (a) The stability condition for the explicit scheme is $s \equiv D\Delta t/\Delta x^2 \leqslant 0.5$.
   (b) The fully implicit scheme is unconditionally stable.
   (c) The weighted scheme is stable independent on $s \equiv D\Delta t/\Delta x^2$ for $\theta \geqslant 0.5$. For $0 < \theta < 0.5$ the stability condition is $s \leqslant 0.5/(1 - 2\theta)$.
   (d) The Crank–Nicolson method is unconditionally stable. The truncation error in this case is of the order $O(\Delta x^2) + O(\Delta t^2)$.
2. Derive the three-point problem of the form (2.32) corresponding to the weighted method.
3. Use the weighted scheme to find the numerical solution of the time-dependent diffusion problem

   $$u_t - \varepsilon u_{xx} = 0 \quad (0 < x < 1,\ t > 0)$$

   subject to the boundary conditions $u(0, t) = \psi_0$, $u(1, t) = \psi_1$ $(t > 0)$ and the initial condition $u(x, 0) = \sin(\pi k x) + (1 - x)\psi_0 + x\psi_1$ $(0 < x < 1)$. Here, ψ_0 and ψ_1 are constants and k is any positive integer. [Analytic solution: $u(x, t) = \sin(\pi k x)e^{-\pi^2 k^2 \varepsilon t} + (1 - x)\psi_0 + x\psi_1$]
4. Find the numerical solution of the time-dependent version of the diffusion problem 2 in exercise 6. Use the weighted scheme. Define the diffusion coefficient $D = 0.02\ \mathrm{m^2\,s^{-1}}$, the length of the computational domain $d = 1$ cm, the source term $S = a\exp[-(x - b)^2/c^2]$ with $a = 4.6 \times 10^{23}\ \mathrm{m^{-3}\,s^{-1}}$, $b = 0.8$ cm, and $c = 0.1$ cm. Use the boundary conditions $n(0, t) = 0$ and $n_x(d, t) = 0$, and the initial condition $n(x, 0) = 10^9\ \mathrm{cm^{-3}}$. Snapshots of the numerical solution at $t = 0, 2, 4, 6, 8$, and 10 ms are shown in figure 2.5. Note that at $t = 10$ ms the solution is at steady state and hence it is identical to the solution of corresponding stationary problem 2 in exercise 6.

2.2.4 Time-dependent convection and diffusion equation

The time-dependent convection–diffusion equation (the *transport* equation) in one space dimension has the form

$$\frac{\partial n}{\partial t} + \frac{\partial}{\partial x}\left(vn - D\frac{\partial n}{\partial x}\right) = S, \tag{2.64}$$

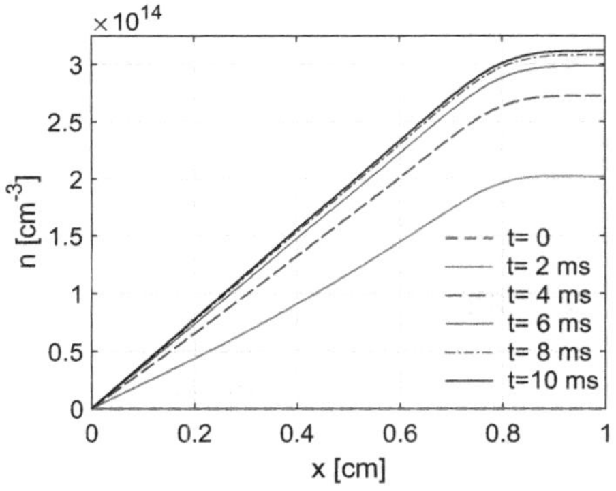

Figure 2.5. Solution of problem 4 from exercise 8.

or

$$\frac{\partial n}{\partial t} + \frac{\partial \Gamma}{\partial x} = S, \tag{2.65}$$

where $\Gamma = vn - D\frac{\partial n}{\partial x}$ denotes the total flux density. Let us integrate equation (2.64) over the control volume (see figure 2.2), and also over the time interval from some t to $t + \Delta t$:

$$\int_t^{t+\Delta t}\int_{x_w}^{x_e}\left(\frac{\partial n}{\partial t}\right)dtdx + \int_t^{t+\Delta t}\int_{x_w}^{x_e}\left(\frac{\partial \Gamma}{\partial x}\right)dtdx = \int_t^{t+\Delta t}\int_{x_w}^{x_e} S\ dtdx. \tag{2.66}$$

The right-hand side of this equation is approximated by the value of the source term S at grid point P times the volume size Δx and the time step Δt,

$$\int_t^{t+\Delta t}\int_{x_w}^{x_e} S\ dtd = S_P \Delta t \Delta x.$$

The first term in the left-hand side is approximated as

$$\int_{x_w}^{x_e}\left(\int_t^{t+\Delta t}\frac{\partial n}{\partial t}dt\right)dx = \int_{x_w}^{x_e}(n|_{t+\Delta t} - n|_t)dx = (n_P^1 - n_P^0)\Delta x,$$

where the superscript 0 indicates value of the density n at the current time level and superscripts 1 at the next time level.

The actual form of the numerical scheme depends on the approximation of fluxes in the second term of equation (2.66) over time and space axes. By weighting values of fluxes in this term at the current and new time levels,

$$\int_t^{t+\Delta t}\left(\int_{x_w}^{x_e}\frac{\partial \Gamma}{\partial x}dx\right)dt = \int_t^{t+\Delta t}(\Gamma|_{x_e} - \Gamma|_{x_w})dt = [\theta\,(\Gamma_e - \Gamma_w)^1 + (1-\theta)\,(\Gamma_e - \Gamma_w)^0]\Delta t,$$

where $\Gamma_e = \Gamma|_{x_e}$ and $\Gamma_w = \Gamma|_{x_w}$, and the weighting factor θ varies between $0 \leqslant \theta \leqslant 1$, and then putting this into equation (2.66), we get

$$\frac{n_P^1 - n_P^0}{\Delta t}\Delta x + \theta(\Gamma_e - \Gamma_w)^1 + (1 - \theta)(\Gamma_e - \Gamma_w)^0 = S_P\Delta x. \tag{2.67}$$

As can be noted, this method incorporates both fully implicit and explicit methods. Indeed, with $\theta = 0$ this scheme is equivalent to the explicit method, while with $\theta = 1$ this is fully implicit method. In the following, in order to summarize the approach briefly, we consider the case of fully implicit method setting $\theta = 1$, and also drop the superscript 1 in the equations.

The total flux through the *east* face x_e of the control volume is

$$\Gamma_e = \left(vn - D\frac{\partial n}{\partial x}\right)\Bigg|_{x_e} = (vn)|_{x_e} - D_e\frac{n_E - n_P}{(\delta x)_e} = (vn)_e - G_e(n_E - n_P), \tag{2.68}$$

where n_e and v_e stand for the density and convection velocity at east face of the cell, and $G_e = D_e/(\delta x)_e$ is the diffusion conductance. It should be noted that, whereas G_e always remains positive, the convection flux $(vn)_e$ can take either positive or negative values depending on the direction of the convection flow.

Approximation of the diffusive flux $-D\partial n/\partial x$ at x_e in equation (2.68) corresponds to the central difference formula. The principal factor toward formulating the numerical scheme is related to approximation of the convection term vn. Assuming a piecewise-linear profile for the density n, its value at interface x_e becomes equal to the mean of n_E and n_P, $0.5(n_E + n_P)$, which leads to the central difference scheme. Let us consider the upwind method, according to which value of the density n at an interface is equal to n at the grid point on the upwind side of the face,

$$n_e = \begin{cases} n_P & \text{if} \quad v_e > 0, \\ n_E & \text{if} \quad v_e < 0. \end{cases}$$

Denoting $[a, b] = \max(a, b)$, the greater of a and b, the convection flux vn at point e becomes

$$(vn)_e = n_P[v_e, 0] - n_E[-v_e, 0].$$

Then, the total flux density at the interface x_e in equation (2.68) obtains the form

$$\begin{aligned}\Gamma_e &= n_P[v_e, 0] - n_E[-v_e, 0] - G_e(n_E - n_P) \\ &= n_P(G_e + [v_e, 0]) - n_E(G_e + [-v_e, 0]).\end{aligned} \tag{2.69}$$

The value of Γ_w in equation (2.67) is defined similarly:

$$\begin{aligned}\Gamma_w &= \left(vn - D\frac{\partial n}{\partial x}\right)\Bigg|_{x_w} = (vn)|_{x_w} - D_w\frac{n_P - n_W}{(\delta x)_w} \\ &= n_w v_w - G_w(n_P - n_W) \\ &= n_W[v_w, 0] - n_P[-v_w, 0] - G_w(n_P - n_W) \\ &= -n_P(G_w + [-v_w, 0]) + n_W(G_w + [v_w, 0]),\end{aligned} \tag{2.70}$$

where $G_w = D_w/(\delta x)_w$. Putting (2.69) and (2.70) in equation (2.67), we obtain eventually

$$a_P n_P = a_E n_E + a_W n_W + b, \tag{2.71}$$

where

$$a_E = G_e + [-v_e, 0],\ a_W = G_w + [v_w, 0], \tag{2.72}$$

$$a_P = G_e + [v_e, 0] + G_w + [-v_w, 0] = a_E + a_W + (v_e - v_w) + \frac{\Delta x}{\Delta t}, \tag{2.73}$$

$$b = S_P \Delta x + n_P^0 \frac{\Delta x}{\Delta t}. \tag{2.74}$$

Exercise 9 (Time-dependent convection–diffusion equation)

1. Derive the three-point problems of the form (2.32) for equation (2.64), based on the weighted method for progression in time, and approximating convection flux according to
   (a) central differences scheme;
   (b) upwind method;
   (c) exponential scheme.
2. Explore the *consistency*, *stability*, and *convergence* of the numerical methods for equation (2.64). Consider the homogeneous equation ($S \equiv 0$), constant D and v, and define $s = D\Delta t/\Delta x^2$ and the Courant number $C = v\Delta t/\Delta x$. (The stability can be studied by the Von Neumann method, see e.g., [3].) Show that
   (a) with the explicit scheme ($\theta = 0$ in equation (2.67)), and the central differences method for the convection flux, the stability condition is $0 \leqslant C^2 \leqslant 2s \leqslant 1$;
   (b) with the explicit scheme ($\theta = 0$), and the upwind method for the convection flux, the stability condition is $C + 2s \leqslant 1$;
   (c) with the fully implicit scheme ($\theta = 1$), and the upwind method for the convection flux, the algorithm is unconditionally stable;
   (d) with the Crank–Nicolson scheme ($\theta = 0.5$), the algorithm is unconditionally stable.
3. Find the numerical solution of the time-dependent convection–diffusion problem

   $$\frac{\partial u}{\partial t} + \frac{\partial}{\partial x}\left(xu - \varepsilon\frac{\partial u}{\partial x}\right) = f(x, t) \quad (-1 < x < 1, t > 0)$$

   subject to the boundary conditions $u(-1, t) = u(1, t) = -\cos(2t)$ $(t > 0)$ and the initial condition $u(x, 0) = \cos(\pi x)$ $(-1 < x < 1)$. Use

   $$f(x, t) = -2\sin(2t)\cos(\pi x) + \cos(2t)[(1 + \varepsilon\pi^2)\cos(\pi x) + \pi x \sin(\pi x)]$$

apply the methods formulated in problem 1.
[Analytic solution: $u(x, t) = \cos(2t)\cos(\pi x)$]

4. Find the numerical solution of the drift–diffusion problem (2.64) for positive ions. Apply the methods formulated in problem 1. Define the mobility and diffusion coefficients by $\mu p = 0.80$ m^2 Torr/(sV) and $D = \mu k_B T/e$. The source term $S = a\exp[-(x-b)^2/c^2]$ with $a = 4.6\times10^{23}$ m^{-3} s^{-1}, $b = 0.8$ cm, and $c = 0.1$ cm. The electric field strength $E = 3.4\exp(675x)$ V/m. Set the length of the computational domain $d = 1$ cm, the pressure $p = 1$ Torr, and the temperature $T = 290$K. Apply the boundary conditions $n(0, t) = 0$ and $n_x(d, t) = 0$, and the initial condition $n(x, 0) = 10^6$ cm^{-3}. Snapshots of the numerical solution at $t = 0, 2, 4, 6, 8$, and $10\ \mu s$ are shown in figure 2.6. Note that at $t = 10\ \mu s$ the solution is at steady state and hence it is identical to the solution of corresponding stationary problem 4 in exercise 7.

2.3 Fluid models for gas discharge

The reliability of any model is determined by its bottleneck, i.e., by the least accurately known element. In fluid models of gas discharges, such an element is the treatment of electron dynamics. The reason is that the electrons in these plasmas are generally far from equilibrium: electron transport in gas discharges is highly non-local [5, 6]. Exact description of the electron dynamics in gas discharges can be obtained from the numerical solution of the electron Boltzmann equation [7–10]. Another way to improve the drawbacks of fluid models is provided by hybrid models, where slow plasma species are treated within the framework of fluid model, while fast species (in particular, fast electrons, usually by Monte Carlo simulations) are treated as particles [11–16].

In fluid models of gas discharges, the background gas is usually treated as a motionless homogeneous medium; all the plasma species, including electrons, are considered as interacting fluid components transported through this neutral gas background. For each plasma species, these models involve the continuity, momentum, and energy equations (usually only for electrons), which are velocity moments

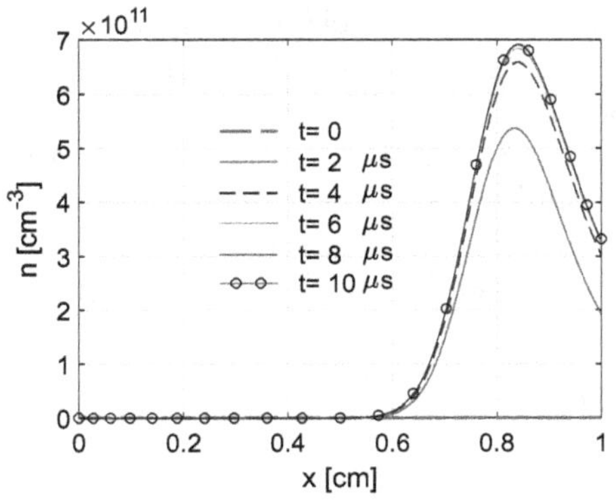

Figure 2.6. Solution of problem 4 from exercise 9.

of the kinetic Boltzmann equation (see chapter 1). As was described in section 1.4, the continuity and momentum equations in these models are usually reduced to the drift–diffusion equations. The effect of particle collisions is taken into account through the transport and reaction rate coefficients, which constitute the input data for the fluid model [17, 18]. Thus, the gas discharge model includes equations of the form (1.41)

$$\frac{\partial n_k}{\partial t} + \nabla \cdot \mathbf{\Gamma}_k = S_k \tag{2.75}$$

for charged and excited species, with the particle flux density $\mathbf{\Gamma}$ in the drift–diffusion approximation,

$$\mathbf{\Gamma}_k = sgn(q_k) n_k \mu_k \mathbf{E} - D_k \nabla n_k. \tag{2.76}$$

Electric field is determined from the solution of Poisson's equation in the electrostatic approximation

$$\varepsilon_0 \nabla \cdot \mathbf{E} = \sum_k q_k n_k, \quad \mathbf{E} = -\nabla\varphi. \tag{2.77}$$

In these equations, ε_0 is the dielectric constant, q is the charge, n is the number density, $\mathbf{E}$ and φ are the electric field and potential, μ and D are the particle mobility and diffusion coefficients, S is the particle creation/destruction rate. Subscripts k indicates the kth species (in the following the subscripts i, e, and g are used to denote the ions, electrons, and background gas particles, respectively).

Within fluid models, a large number of species can be included allowing the study of complicated plasma chemistry involving numerous reactions. Moreover, since the computational speed of these models is relatively high, even two- and three-dimensional fluid simulations can be run in a reasonably short time. These advantages have made fluid models widely used in the numerical simulation of low-temperature plasma discharges.

2.3.1 *Simple* fluid model

Within the *simple* fluid model, the particle mobility and diffusion coefficients as well as the particle creation rates are defined as functions of the reduced electric field E/p or E/N (local field approximation, LFA). The mobility and diffusion are related by the Einstein relation, $D/\mu = k_B T/e$. Here, $p = n\, k_B T$ denotes the gas pressure, T is the kinetic temperature, and k_B is the Boltzmann constant.

In particular, for a two-component plasma composed of electrons and positive ions with number densities n_e and n_i, drift–diffusion equations have the form

$$\frac{\partial n_e}{\partial t} + \nabla \cdot (-\mu_e n_e \mathbf{E} - D_e \nabla n_e) = S_e, \tag{2.78}$$

$$\frac{\partial n_i}{\partial t} + \nabla \cdot (\mu_i n_i \mathbf{E} - D_i \nabla n_i) = S_i. \tag{2.79}$$

The system is supplied with the Poisson's equation for the electric field,

$$\nabla \cdot \mathbf{E} = \frac{e}{\varepsilon_0}(n_i - n_e), \quad \mathbf{E} = -\nabla\varphi. \tag{2.80}$$

The source terms in the equations (2.78) and (2.79) are usually defined by using Townsend ionization coefficient $\alpha(|\mathbf{E}|/p)$ obtained from the experiment

$$S_e = S_i = |\mathbf{\Gamma}_e|\, \alpha(|\mathbf{E}|/p) \tag{2.81}$$

or, alternatively, according with the classical Townsend approximation with $\alpha(|\mathbf{E}|/p) = Ap \exp(-pB/|\mathbf{E}|)$, where A and B are constants determined by the type of gas [17].

The model is completed by definition of boundary and initial conditions. Consider the case of axially symmetric DC discharge with a parallel-plate configuration (see figure 2.7), where the radial coordinate r defines the direction parallel to the plane electrodes, and the axial coordinate z the direction normal to them. Let the anode locate at $z = 0$, the cathode at $z = d$, and the lateral boundary is at $r = l$. Then, the boundary conditions can be defined in the following way. On the anode, $z = 0$, we set

$$n_i = 0, \quad \frac{\partial n_e}{\partial z} = 0, \tag{2.82}$$

and on cathode $z = d$

$$\frac{\partial n_i}{\partial z} = 0, \quad \mu_e n_e = \gamma \mu_i n_i, \tag{2.83}$$

where the second condition describes the so-called γ-process, $\mathbf{\Gamma}_e|_{z=d} \cdot \hat{\mathbf{n}} = -\gamma \mathbf{\Gamma}_i|_{z=d} \cdot \hat{\mathbf{n}}$, where γ is the secondary electron emission coefficient, $\hat{\mathbf{n}}$ is the normal unit vector directed outward from the volume, and the diffusion transport is neglected. Denoting applied DC voltage as U, we specify

$$\varphi(r, 0, t) = U, \quad \varphi(r, d, t) = 0. \tag{2.84}$$

Finally, on the symmetry axis and the lateral boundary of the domain, $r = l$, we set

$$\frac{\partial n_e}{\partial r} = \frac{\partial n_i}{\partial r} = \frac{\partial \varphi}{\partial r} = 0. \tag{2.85}$$

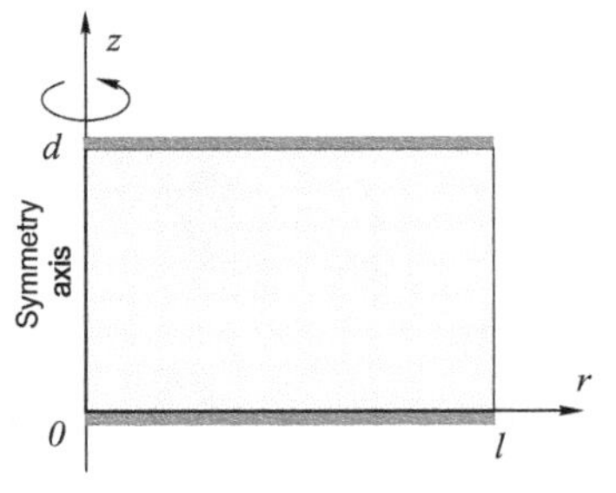

Figure 2.7. Schematic representation of the planar gas discharge cell.

As was noted in the introduction, the bottleneck of the *simple* fluid model is the local approximation (LFA) for the electron energy distribution function (EEDF), according to which the characteristics of processes involving electrons are determined by the local value of the electric field strength at a given point in space and at a given instant in time. In fact, in spatially inhomogeneous fields, for example, in the near-electrode regions of the discharges, the LFA is unsuitable for studying the processes that involve electrons. Let us illustrate this by the following example for near-cathode region of a classical DC glow discharge.

The classical theory describing the physical phenomena in the electrode regions of the gas discharge has been developed by A V Engel and M Steenbeck. It is based on the assumption that the ionization rate depends on the local electric field strength (LFA). The condition for maintaining the discharge in this situation reduces to the Townsend breakdown criterion, with the only exception that the length of the discharge gap is replaced by the thickness of the cathode sheath (see, e.g., [17]). The cathode sheath is assumed to be autonomous in the sense that all processes responsible for maintaining the discharge occur precisely in it.

Under these conditions, the cathode sheath is inevitably followed by the positive column (PC), such that the border between the cathode sheath and plasma coincides with the border between the negative glow (NG) and positive column, and hence the Faraday dark space (FDS) is absent. The ionization (and glow) in this situation concentrates in the cathode sheath, where the electric field is sufficiently strong. Such a discharge structure is in obvious conflict with observations, which suggest that ionization in the cathode region is non-local, that means it is not defined by the local value of the field strength. Indeed, the electrons emitted from the cathode and also those generated in the cathode sheath and accelerated by the strong electric field in it are responsible for non-local ionization in the plasma regions, such as NG, where the field is weak. Therefore, when considering problems of this type, the models based on the LFA can lead to *a priori* physically incorrect results [6].

Exercise 10 (*Simple* fluid model)

1. (**Project Problem**: *1D model for DC glow discharge in argon*) In 1D approximation, equations (2.78)–(2.80) describing a *simple* model for two-component plasma of DC discharge, confined to a single axis normal to the plane electrodes, obtains the form

$$\frac{\partial n_e}{\partial t} + \frac{\partial}{\partial z}(-\mu_e n_e E - D_e\frac{\partial n_e)}{\partial z} = S_e, \tag{2.86}$$

$$\frac{\partial n_i}{\partial t} + \frac{\partial}{\partial z}(\mu_i n_i E - D_i\frac{\partial n_i}{\partial z}) = S_i, \tag{2.87}$$

$$-\frac{\partial^2 \varphi}{\partial z^2} = \frac{e}{\varepsilon_0}(n_i - n_e). \tag{2.88}$$

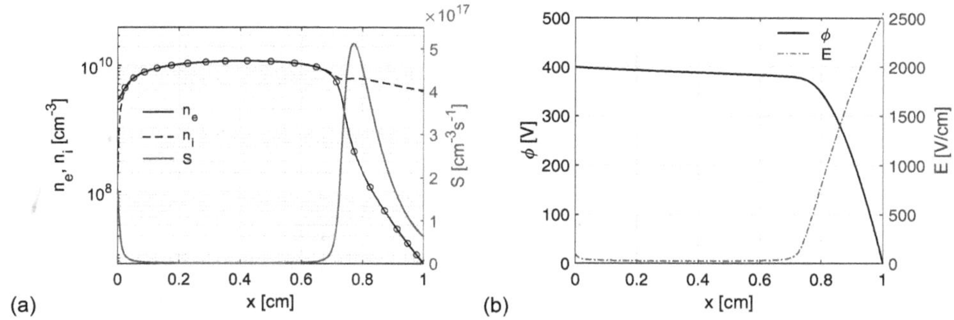

Figure 2.8. Solution of problems 1 and 2 from exercise 10. (a) Electron and ion densities n_e and n_i, and the ionization function S; (b) electric field strength E and potential φ.

Here, n is the particle number density, $E = -\partial\varphi/\partial z$ and φ are the electric field and potential, μ and D are the mobility and diffusion coefficients, subscripts i and e indicate the positive ions and electrons.

Find the numerical solution of this system, subject to following conditions. Set the ionization terms $S_e = S_i = Ap\,|\Gamma_e|\exp(-pB/|E|)$, where Γ_e is the electron flux density, $A = 12$ cm^{-1} Torr^{-1} and $B = 180$ V cm^{-1} Torr^{-1} for argon plasma. Define the mobility and diffusion coefficients by $\mu_i p = 0.80$ m^2 Torr/(sV) and $D_i = \mu_i k_B T_i/e$, $\mu_e p = 30$ m^2 Torr/(sV) and $D_e = \mu_e k_B T_e/e$. Set the length of the computational domain $d = 1$ cm, the pressure $p = 1$ Torr, the temperatures $T_i = 0.025$ eV and $T_e = 1$ eV, and the feeding voltage $U = 400$ V. Adopt the boundary conditions (2.82)–(2.85) to given 1D geometry. Note that computed profiles of the gas discharge characteristics are shown in figure 2.8.

2. (**Project Problem**: *COMSOL Multiphysics implementation of 1D model for DC glow discharge in argon*) COMSOL Multiphysics [19] is finite element computational package. Important advantage of COMSOL Multiphysics is its flexibility in specifying the model, modifying the predefined physics interfaces, etc. The approach given below is based on the **Convection–Diffusion equation** interface.
    - (i) From the **File** menu, choose **New**. In the **New** window, click **Model Wizard**, and select **1D**. The **Select Physics** tree will appear.
    - (ii) In the **Select Physics** tree, from the **Mathematics / Classical PDEs** select the **Stabilized Convection–Diffusion equation** and click **Add**. Define the dependent variable name `ne`. Repeat this step and define the dependent variable name `ni`.
    - (iii) In the **Select Physics** tree, from **Mathematics / Classical PDEs** select **Poisson's equation** interface. Define the dependent variable name `phi`.
    - (iv) Click **Study**. In the **Study** tree, select **Time Dependent**. Click **Done**. COMSOL's main window will be opened.
    - (v) In the **Model Builder** window, under **Global Definitions**, click **Parameters**, and enter the following:

e	1.6e-19	[C]	Electron charge
eps0	8.854e-12	[F m^{-1}]	Vacuum permittivity
gamma	0.07		Secondary emission coefficient
p	1	[Torr]	Pressure
U	400	[V]	Applied voltage
Te	1	[eV]	Electron temperature
Ti	0.025	[eV]	Ion temperature
mue	30	[m^2/(sV)]	Electron mobility
De	mue*Te	[m^2/(s)]	Electron diffusion
mui	0.8	[m^2/(sV)]	Ion mobility
Di	mui*Tg	[m^2/(s)]	Ion diffusion
L	0.01	[m]	Discharge gap length
A	1200	[m^{-1} $Torr^{-1}$]	Coefficient in Townsend formula
B	18000	[V m^{-1} $Torr^{-1}$]	Coefficient in Townsend formula

(vi) In the **Model Builder** window, under the **Component**, from the **Geometry** toolbar select the **Interval** and enter its length `L`.

(vii) Under the **Component/Definitions** by right click select the **Variables**, and enter the following:

E	-phix	Electric field strength
fluxe	-De*nex-ne*mue*E	Electron flux density
fluxi	-Di*nix+ni*mui*E	Ion flux density
Se	abs(fluxe)*alpha	Electron creation (destruction) rate
Si	abs(fluxe)*alpha	Ion creation (destruction) rate
alpha	A*p*exp(-B*p/abs(E))	Townsend coefficient
J	e*(fluxi-fluxe)	Electric current density

From the **Stabilized Convection–Diffusion Equation** module for electrons:

(viii) Select **Convection–Diffusion equation**, and in the **Settings** menu type `De`, `-mue*E`, and `Se` into the fields **Diffusion Coefficient**, **Conservative Flux Convection Coefficient**, and **Source Term**, respectively.

(ix) In the **Settings** menu select **Initial Values** and enter `1e13`.

(x) By right clicking, choose **Dirichlet Boundary Condition**, select the right end point and introduce `gamma*(mui/mue)*ni`.

(xi) Repeat the steps from (viii) to (x) and fill in appropriate fields for the ion transport.

In the **Poisson's equation** module **Settings**:

(xii) Enter `(e/eps0)*(ni-ne)` into the **Source Term**.

(xiii) Select **Initial Values** and enter `U*(1-x/L)`.
(xiv) By right clicking, select **Dirichlet Boundary Condition**, and set `U` at the left end point (the anode). Repeat and define boundary condition 0 at the right end point (the cathode).
(xv) By right clicking **Mesh**, select **Edge/Distribution** and introduce number of grid points.
(xvi) From the **Model Wizard** select **Study/Solver Configurations/Show Default Solver**.
(xvii) Under **Study/Solver Configurations/Solution/ Time-Dependent Solver** set the time integration interval from 0 to 1e-5, **Initial step** 1e-12 and **Maximum step** 1e-6.
(xviii) Click **Compute**.
(xix) Post processing is carried out in **Results** in the **Model Wizard**. Note that the gas discharge characteristics computed under given parameters are shown in figure 2.7.

2.3.2 *Extended* fluid model

The *simple* fluid model for glow discharges (section 2.3.1) seems be capable of describing the basic properties of the gas discharge in a self-consistent way. However, it turns out that this model is unable to account for the ionization properly in the discharge regions, where the electric field is weak and the ionization is non-local.

An approach called an *extended fluid model*, also known as a *local mean energy approximation*, LMEA, as opposed to LFA, intends to eliminate the shortcomings of the LFA by incorporating, to a certain extent, non-local transport of electrons into fluid models. It was suggested in [20] and its different modifications were applied in [14, 21–27]. The idea behind this method is that the electron transport and kinetic coefficients, which are the electron mobility and diffusion coefficients as well as the electron-impact reaction rate constants for excitation and ionization processes, are determined from the solution of the kinetic Boltzmann equation as functions of the electron temperature T_e rather than the local value of the reduced electric field, E/N. Within this model, profile of the electron temperature, T_e, is obtained from the solution of the energy balance equation for electrons, which along with the volume processes includes the transfer by heat conduction. The electron heating source in this situation becomes non-local, because the effect of heating caused by the heat flux from the discharge regions where the Joule heating basically occurs (generally from the near-electrode sheath) is taken into account.

Equations

In order to incorporate the non-local transport of electrons into the fluid model for gas discharge, the set of equations consisting of the drift–diffusion equations for particle balance (see, e.g., [23]),

$$\frac{\partial n_k}{\partial t} + \nabla \cdot \mathbf{\Gamma}_k = S_k, \qquad (2.89)$$

where $\boldsymbol{\Gamma}$ is the particle flux density,

$$\boldsymbol{\Gamma}_k = sgn(q_k)n_k\mu_k\mathbf{E} - D_k\nabla n_k, \tag{2.90}$$

and the Poisson's equation for the electric field,

$$\varepsilon_0\nabla \cdot \mathbf{E} = \sum_k q_k n_k, \quad \mathbf{E} = -\nabla\varphi. \tag{2.91}$$

is supplied with the electron energy equation

$$\frac{\partial n_\varepsilon}{\partial t} + \nabla \cdot \boldsymbol{\Gamma}_\varepsilon = -e\,\boldsymbol{\Gamma}_e \cdot \mathbf{E} - \frac{3}{2}\frac{m_e}{m_g}\nu_{e,a}n_e k_{\mathrm{B}}\left(T_e - T_g\right) - \sum_j \Delta E_j R_j. \tag{2.92}$$

In these equation, $n_\varepsilon = n_e\bar{\varepsilon}$ is the electron energy density, $\bar{\varepsilon} = 3/2k_{\mathrm{B}}T_e$ denotes the mean electron energy, and the density of the electron energy flux is

$$\boldsymbol{\Gamma}_\varepsilon = -D_\varepsilon\nabla n_\varepsilon - \mu_\varepsilon\mathbf{E}n_\varepsilon. \tag{2.93}$$

In the right-hand side of equation (2.92), the first term defines the Joule heating or cooling of electrons by the electric field, and the second and third terms describe the energy loss due to elastic and inelastic collisions, correspondingly. In this equation, $\nu_{e,a}$ is the electron-atomic elastic collision frequency, ΔE_j and R_j are the energy loss or gain due to inelastic collision and corresponding reaction rate, T_g is the background gas temperature, m is the particle mass.

Transport coefficients

Let us illustrate this model for the case of plasma of DC gas discharge sustained in argon gas. The three species, namely, electrons, positive ions, and metastable atoms are taken into consideration. The set of reactions is listed in table 2.2, and collision cross-sections are depicted in figure 2.9. The first process in table 2.2 describes the elastic scattering of electrons. The effective frequency, $\nu_{e,a}$, mobility, μ_e, and diffusion, D_e, coefficients are calculated using the cross-section for this process,

$$\mu_e = -\frac{1}{n_e}\frac{e}{m_e}\int_0^\infty D_r\sqrt{\varepsilon}\frac{\partial}{\partial\varepsilon}f_0(\varepsilon)d\varepsilon, \tag{2.94}$$

$$D_e = \frac{1}{n_e}\int_0^\infty D_r\sqrt{\varepsilon}f_0(\varepsilon)d\varepsilon, \tag{2.95}$$

where $\varepsilon = mv^2/2e$ is the electron kinetic energy (in eV units), $D_r = 2\varepsilon/3m_e\nu_{e,a}$ is the space diffusion coefficient, $f_0(\varepsilon)$ is the EEDF obtained from the solution of local Boltzmann equation, and satisfying the normalization condition

$$\int_0^\infty f_0(\varepsilon)\sqrt{\varepsilon}\,d\varepsilon = 1. \tag{2.96}$$

Heavy particle are assumed to follow the Maxwell distribution function. Their mobility and diffusion coefficients are defined by constant parameters, which depend

Table 2.2. The set of elementary reactions. Label Boltz. indicates that the rate constant was calculated from local Boltzmann equation.

Index	Reaction	Type	ΔE(eV)	Constant
1	$e + Ar \rightarrow e + Ar$	Elastic collision	0	Boltz.
2	$e + Ar \rightarrow 2e + Ar^{+}$	Direct ionization	15.8	Boltz.
3	$e + Ar \leftrightarrow e + Ar^{*}$	Excitation	11.4	Boltz.
4	$e + Ar \rightarrow e + Ar$	Excitation	13.1	Boltz.
5	$e + Ar^{*} \rightarrow 2e + Ar^{+}$	Stepwise ionization	4.4	Boltz.
6	$2Ar^{*} \rightarrow e + Ar^{+} + Ar$	Penning ionization	—	6.2×10^{-10} cm^{-3} s^{-1}
7	$Ar^{*} \rightarrow h\nu + Ar$	Radiation (including trapping)	—	1.0×10^{7} s^{-1}

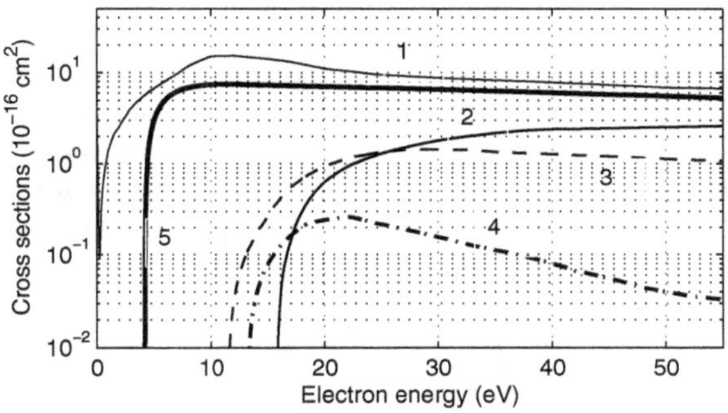

Figure 2.9. Electron cross-sections for (1) elastic, (2) direct ionization, (3) excitation, (4) excitation, and (5) stepwise ionization collisions in argon, used in the model. Curve labels correspond to indices of corresponding processes in table 2.2. Adapted from [46] with permission of AIP Publishing.

on the background gas density. Moreover, the ion mobility and diffusion are related by the Einstein relation $D_i/\mu_i = k_B T_i/e$ with $T_i = T_g$.

Energy transport coefficients in (2.93) are related to particle transport coefficients through $\mu_\varepsilon = (5/3)\mu_e$ and $D_\varepsilon = (5/3)D_e$ [23]. Note that the electron energy flux $\mathbf{\Gamma}_\varepsilon$ (2.93) can also be expressed in the form

$$\mathbf{\Gamma}_\varepsilon = \frac{5}{2} k_B T_e \mathbf{\Gamma}_e - \lambda_e \nabla k_B T_e, \tag{2.97}$$

with $\lambda_e = \frac{5}{2} n_e D_e$ [21, 23].

Source terms

Volume source terms S_k in the particle balance equations (2.75) are determined by the reactions happening in the discharge,

$$S_k = \sum_i R_i - \sum_j R'_j, \tag{2.98}$$

where the rates of corresponding reactions R_i and R'_j are proportional to the densities of the reacting components.

The direct, stepwise, and Penning ionization processes determine the electron balance in the model,

$$S_e = S_i = R_2 + R_5 + R_6 = K_2 n_0 n_e + K_5 n_m n_e + K_6 n_m^2, \tag{2.99}$$

where, due to particle conservation, the reaction rates for electrons and ions are equal. Here, R_2, R_5, and R_6 stand for the direct, stepwise, and Penning ionization reaction rates, K_2, K_5, and K_6 are the constants of these reactions, and n_0 denotes the density of neutral atoms. Here, the indices of reaction rates correspond to those of the processes in table 2.2.

The reactions of excitation, stepwise ionization, Penning ionization, and radiation regulate the balance of excited atoms,

$$S_m = R_3 - R_5 - 2R_6 - R_7 = K_3 n_0 n_e - K_5 n_m n_e - 2K_6 n_m^2 - K_7 n_m. \tag{2.100}$$

The reaction rate constant K depends on the energy distribution function of the related component. For electron-induced reactions (processes 1–5 in table 2.2), the rate constants of these processes are computed by convolving the EEDF, obtained from solution of local Boltzmann kinetic equation, with corresponding cross-sections,

$$K_R = \int_0^\infty \sigma_R(\varepsilon)\sqrt{\varepsilon}\, f_0(\varepsilon) d\varepsilon. \tag{2.101}$$

Dependence of collision cross-sections σ_R of reaction R on energy ε is depicted in figure 2.9.

The so-called look-up-tables (LUT's) that relate the mean energy (temperature) of electrons to their kinetic coefficients, such as the electron transport (diffusion, D_e, and mobility, μ_e) coefficients as well as the rate constants K_R of electron-induced plasma-chemical reactions for excitation and ionization are usually computed using a separate electron Boltzmann equation solver [28]. Incorporating the LUT's into the numerical model significantly reduces the processing time because in this situation computationally expensive Boltzmann solver runs only once.

Boundary conditions

Let us consider the case of axially symmetric DC discharge sustained in cylindrical tube with parallel-plate electrodes (see figure 2.7). Boundary condition for the ions and metastable atoms at the electrodes and the wall of the discharge tube are defined in the following way (see, e.g., [20, 29, 30]):

$$\hat{\mathbf{n}} \cdot \boldsymbol{\Gamma}_i = 1/4 v_i n_i + \alpha n_i \mu_i (\hat{\mathbf{n}} \cdot \mathbf{E}), \tag{2.102}$$

$$\hat{\mathbf{n}} \cdot \boldsymbol{\Gamma}_m = 1/4 v_m n_m. \tag{2.103}$$

Here, $v_j = \sqrt{8 k_B T_j / \pi m_j}$ is the thermal velocity ($j = e, i, m$), the particle flux density $\boldsymbol{\Gamma}$ is defined by equation (2.90), $\hat{\mathbf{n}}$ is the normal unit vector directed outward from the volume, and α is a switching function (either 0 or 1) such that $\alpha = 1$ if $(\hat{\mathbf{n}} \cdot \mathbf{E}) > 0$ at the boundary surface, and $\alpha = 0$ otherwise.

Boundary conditions for the electron number and energy densities at the anode and dielectric wall are defined by equations

$$\hat{\mathbf{n}} \cdot \mathbf{\Gamma}_e = 1/4 v_e n_e, \tag{2.104}$$

$$\hat{\mathbf{n}} \cdot \mathbf{\Gamma}_\varepsilon = 1/3 v_e n_\varepsilon, \tag{2.105}$$

while at the cathode

$$\hat{\mathbf{n}} \cdot \mathbf{\Gamma}_e = 1/4 v_e n_e - \gamma\, \hat{\mathbf{n}} \cdot \mathbf{\Gamma}_i, \tag{2.106}$$

$$\hat{\mathbf{n}} \cdot \mathbf{\Gamma}_\varepsilon = 1/3 v_e n_\varepsilon - 2k_{\mathrm{B}} T_e \gamma\, \hat{\mathbf{n}} \cdot \mathbf{\Gamma}_i, \tag{2.107}$$

where γ is the secondary electron emission coefficient. The electron particle and energy density fluxes at boundaries (2.104), (2.106) and (2.105), (2.107) are related by the equation $\hat{\mathbf{n}} \cdot \mathbf{\Gamma}_\varepsilon = 2k_{\mathrm{B}}\, T_e\, (\hat{\mathbf{n}} \cdot \mathbf{\Gamma}_e)$.

The symmetry condition $\hat{\mathbf{n}} \cdot \mathbf{\Gamma}_j = 0$ is imposed along the z-axis for all variables ($j = i, e, m, n_\varepsilon$).

The electrostatic potential $\varphi = 0$ at the cathode and $\varphi = U_d$ at the anode. At the dielectric surface

$$\frac{\partial \varphi}{\partial r} = \frac{1}{\varepsilon_0} \sigma, \tag{2.108}$$

where the surface charge density σ is calculated from the equation

$$\frac{\partial \sigma}{\partial t} = \hat{\mathbf{n}} \cdot \mathbf{J}, \tag{2.109}$$

where $\mathbf{J} = e(\mathbf{\Gamma}_i - \mathbf{\Gamma}_e)$ is the current density.

Modeling results

Calculations are conducted for the discharge in argon gas. Three chemical species, namely, electrons (e), positive ions (Ar^+), and metastable atoms (Ar^*), and seven elementary reactions, which are the electron-impact ionization, electronic excitation from ground-state, stepwise ionization, Penning ionization, and radiative de-excitation, are considered in gas phase (see table 2.2). The cross-sections of elementary processes are shown in figure 2.9.

Current–voltage characteristic (CVC) curves

CVC curves in figure 2.10 correspond to the discharge regime, where the argon pressure $p = 3$ Torr, the discharge gap $L = 1$ cm, the discharge tube radius $R = 1.5$ cm, and the secondary emission coefficient $\gamma = 0.1$ [21, 31]. Basic parameters of the discharge corresponding to the points indicated along the CVC curve in figure 2.10 are listed in table 2.3. I_d and U_d denote the discharge current and voltage, d is the width of the cathode layer, and r^* is the radius of the cathode spot. The left descending branch of the CVC curve (points 1–3) constitutes the subnormal discharge, the central part (points 3–5) corresponds to the normal, and the right ascending branch (points 5–7) to the abnormal glow discharge. In the subnormal regime, voltage drops from 280 to 220 V while the current varies from 0.01 to 0.6 mA. In the normal glow, the discharge voltage keeps essentially constant (about 220 V) over the current range from 0.6 mA to 10 mA. In the abnormal regime, the voltage grows from 220 to 300 V with currents ranging from 10 mA to 500 mA.

It should be noted that simulations of the discharge in the abnormal mode, which is stable, are carried out without incorporating the external circuit equation for the voltage U_d into the model, specifying the value of U_d instead. However, when simulating the discharge in subnormal and normal modes, the equations (2.75)–(2.92) are completed with the external circuit equation,

$$\frac{dU_d}{dt} + \frac{1}{\mathcal{C}}\left(I_d - \frac{U_{src} - U_d}{\mathcal{R}}\right) = 0. \tag{2.110}$$

Otherwise, the numerical solution converges to the abnormal discharge regime, corresponding to the specified voltage U_d. In this equation, U_{src} denotes the applied DC voltage, $\mathcal{R}$ is the resistance of the circuit, and $\mathcal{C}$ is the external capacitance. Here, $\mathcal{C} = 1$ pF and $U_{src} = 500$ V, and in order for different discharge regimes depicted in figure 2.10 to be described, the resistance $\mathcal{R}$ is varied between 500 Ω and 10 MΩ.

In the normal glow regime, in agreement with experimental observations [17], the area of the cathode spot continuously increases with increasing current, while the current density on the cathode surface remains constant. Figure 2.10(b) demonstrates this inherently 2D characteristic of the discharge, which therefore cannot be realized within the 1D model. It should be emphasized that the horizontal axis in this figure is spanned by the current density J_d rather than the current magnitude I_d, and that the points indicated along the CVC curves in panels (a) and (b) correspond to the same regimes. Note that while points 3, 4, and 5 in panel (a) are separated from one another, that means that the current magnitudes are different for these regimes,

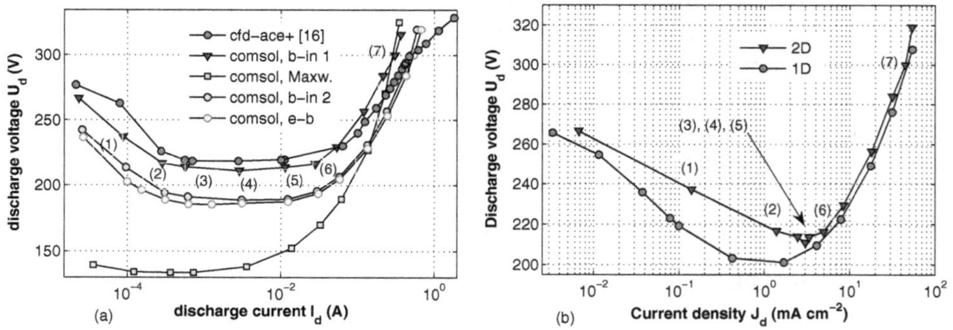

Figure 2.10. The current–voltage characteristic of the discharge in argon at pressure $p = 3$ Torr, the discharge gap $L = 1$ cm, the tube radius $R = 1.5$ cm, and $\gamma = 0.1$. (a) $U_d = U_d\,(I_d)$, (b) $U_d = U_d\,(J_d)$. Adapted from [46] with permission of AIP Publishing.

Table 2.3. Parameters of the glow discharge for points 1–7 along the current–voltage curve in figure 2.10.

	U_d (V)	I_d (mA)	J_d (mA cm^{-2})	r^* (cm)	d (cm)
1	236	0.088	0.14	0.26	0.60
2	214	0.29	1.39	0.15	0.18
3	210	0.58	2.46	0.18	0.14
4	207	2.93	3.03	0.47	0.13
5	208	11.7	3.33	0.97	0.13
6	216	28.4	4.93	1.29	0.11
7	300	310	44.4	1.48	0.058

the same points in panel (b) practically coincide, and hence the current densities for these points are nearly the same.

Typical glow discharge regimes

Let us focus on the analysis of the three typical glow discharge modes. For this purpose, we consider spatial distributions of the discharge properties, corresponding to the regimes indicated by points 1, 4, and 7 in figure 2.10 and table 2.3.

Point 1 corresponds to the subnormal mode. As can be seen from 2D plots of the ion and electron densities (figure 2.11(a) and (b)) and from the axial density profiles (figure 2.11(c)), quasi-neutrality in this regime is not fulfilled. Since the electric current is rather weak and the electron mobility far exceeds that of ions, the ion density is greater than the electron density over the entire discharge volume [17].

Figure 2.12 shows the densities of charged particles for point 4, which corresponds to the normal mode. As is evident from this figure, quasi-neutrality is violated in the cathode layer only, whereas in the remaining part of the discharge the electron and ion number densities are approximately equal, $n_e \approx n_i$.

Finally, figure 2.13 shows 2D and axial density plots for the abnormal discharge that corresponds to point 7. In this regime, as can be seen from the density axial profiles, the cathode layer tends to decrease in width, leaving the rest of the discharge volume quasi-neutral, $n_e \approx n_i$.

Comparison of 1D and 2D modeling results

A significant part of numerical investigations of gas discharges is restricted to one-dimensional geometry. The reason is that 1D models strongly reduce the computational cost, that is especially important in the case of kinetic and particles approaches, which are extremely time consuming. However, it is obvious that accuracy of 1D modeling results, in general, is of limited utility. Here, as a test of the applicability of 1D approximation, we compare results obtained from 1D and 2D models for the discharge conditions considered in the previous section: argon pressure is $p = 3$ Torr, the discharge gap is $L = 1$ cm, and the radius of the discharge tube is $R = 1.5$ cm. Figure 2.10 (b) shows CVC curves $U_d = U_d(J_d)$, where U_d and J_d stand for the discharge voltage and current density, for the 1D and 2D discharge

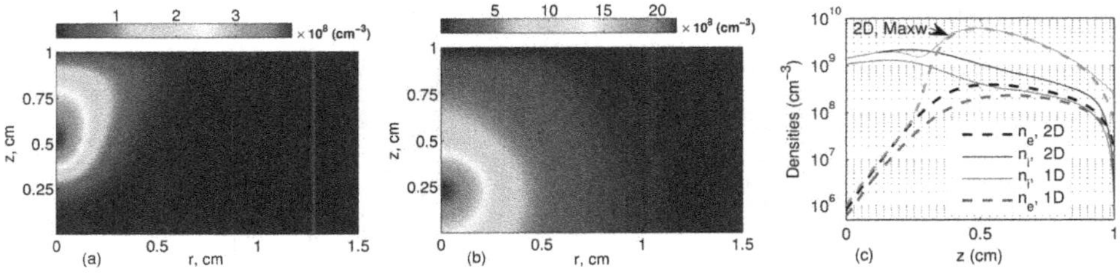

Figure 2.11. 2D profiles of the electron n_e (a) and ion n_i (b) densities, and axial profiles of the electron and ion (c) densities. Blue lines in (c) correspond to 1D solution, black lines to 2D solutions, and red lines to 2D solution with the Maxwellian EEDF. Parameter regime: point 1 in figure 2.10 and table 2.3. Adapted from [46] with permission of AIP Publishing.

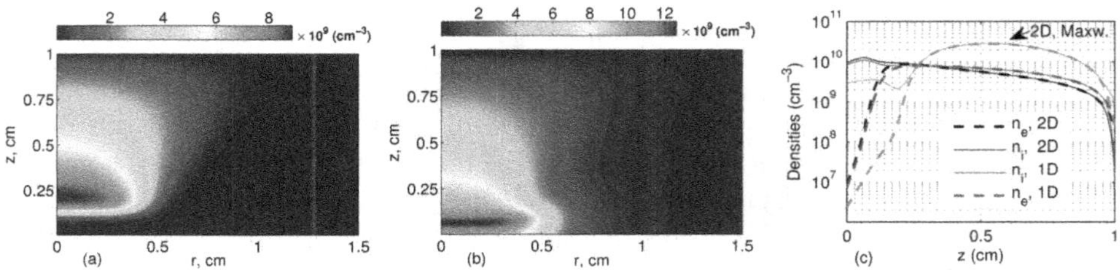

Figure 2.12. The same as in figure 2.11 but for the regime 4 in figure 2.10 and table 2.3. Adapted from [46] with permission of AIP Publishing.

models. Note that CVC for 1D discharge has the form $U_d = U_d(J_d)$ by itself. However, in the 2D case, current density J_d is determined as an average over the cathode spot, with the spot radius equal to the distance from the discharge tube axis where the current density reduces by one half. These current density magnitudes are listed in table 2.3. Values of the normal cathode layer thickness $pd = 0.39$ Torr cm and the normal current density $J_d/p^2 = 0.34$ mA/(cm^2 Torr^2) correlate well with the published data (see, e.g., [17], pp 182–183 and [32]).

Let us consider the axial profiles of the discharge characteristics derived from 1D and 2D models, for the three parameter regimes indicated by points 1, 4, and 7 in figure 2.10 and table 2.3. The discharge currents for theses regimes are $I_d = 8.8 \times 10^{-5}$ A, 2.9×10^{-3} A, and 0.31 A. In order to observe discharges in similar regimes in 1D and 2D models, the conditions for 1D discharges were chosen to obtain current densities fitting the mean (over the cathode spot) current densities of the corresponding 2D discharges.

Panels (c) in figures 2.11–2.13 compare axial profiles of the electron and ion number densities, n_e and n_i, in subnormal, normal, and abnormal regimes, for 1D and 2D models. Axial profiles of the electric potential, φ, and the electron temperature, T_e, derived from the 1D and 2D models are depicted in figure 2.14(a), (b). For the present discharge conditions with $L < R$, as it is evident from these figures, 1D model predicts well the glow discharge properties, especially in the abnormal mode,

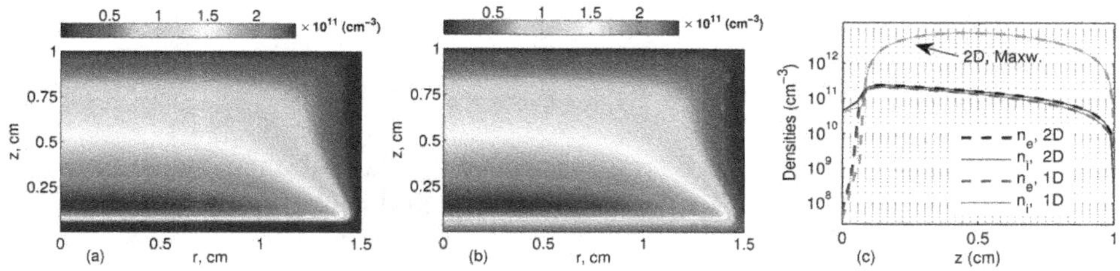

Figure 2.13. The same as in figure 2.11 but for the regime 7 in figure 2.10 and table 2.3. Adapted from [46] with permission of AIP Publishing.

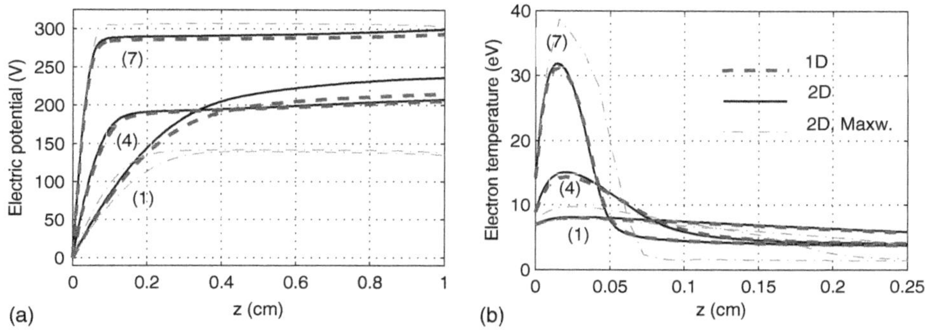

Figure 2.14. Axial profiles of (a) electric potential, φ, and (b) electron temperature, T_e, for the regimes 1, 4, and 7 in figure 2.10 and table 2.3. Dotted lines: 1D solution, solid lines: 2D solution. Red lines correspond to 2D solution with the Maxwellian EEDF. Adapted from [46] with permission of AIP Publishing.

where the spatial profiles of the discharge characteristics are essentially 'one-dimensional'. Note that in figure 2.10(b) abnormal branches of CVC curves for 1D and 2D models are spaced closely to one another.

This agreement between 1D and 2D models is confirmed by the analysis in [25], which states that the 1D approximation for the glow discharge model is appropriate if the transverse dimension of the discharge exceeds the longitudinal dimension, that means that the discharge radius R is greater than the discharge gap L, and the discharge occupies the cathode surface entirely. This agreement is violated in the normal mode due to the 2D effects such as formation of the normal current density and discharge contraction, which cannot be reproduced within the 1D approximation.

Particle flux profiles

Figures 2.15–2.17 show axial and radial profiles of particle fluxes in the discharge, for the typical three regimes. These are the abnormal, normal, and subnormal modes, corresponding to the points 7, 4, and 1 in the CVC curve in figure 2.10. Axial profiles of the electron drift, $\mathbf{\Gamma}_{e,\text{drift}} = -n_e\mu_e\mathbf{E}$, the electron diffusion, $\mathbf{\Gamma}_{e,\text{diff}} = -D_e\nabla n_e$, and of the total electron flux, $\mathbf{\Gamma}_e = -n_e\mu_e\mathbf{E} - D_e\nabla n_e$, as well as axial profiles of the total ion flux $\mathbf{\Gamma}_i = n_i\mu_i\mathbf{E} - D_i\nabla n_i$, and the difference $\mathbf{\Gamma}_e - \mathbf{\Gamma}_i$ of the electron and ion fluxes, which is proportional to the current density, are shown in figure 2.15. Radial profiles of the total electron and ion fluxes $\mathbf{\Gamma}_e$ and $\mathbf{\Gamma}_i$ as well as their difference $\mathbf{\Gamma}_e - \mathbf{\Gamma}_i$ along the discharge cell midsection, which is the line $z = 0.5$ cm, are demonstrated in figure 2.16. As is evident from panels (a) and (b) of this figure, $|\mathbf{\Gamma}_e| \gg |\mathbf{\Gamma}_i|$ throughout the discharge cross-section for the regimes 7 and 4, where the plasma is essentially quasi-neutral. Finally, figure 2.17 depicts the drift and diffusion components $\mathbf{\Gamma}_{e,\mathrm{drift}}$ and $\mathbf{\Gamma}_{e,\mathrm{diff}}$ of the electron flux from figure 2.16, which exhibit the fact that the total electron flux is about one order of magnitude smaller compared to its drift and diffusion components.

In conclusion, recall that correct description of glow discharges requires a kinetic treatment of the electron dynamics. Within the so-called *extended* fluid model considered in this section, electron transport (diffusion and mobility) coefficients as well as the rates of electron-induced plasma-chemical reactions are determined from the solution of the Boltzmann equation for EEDF and corresponding collision cross-sections. Analysis shows that this model describes adequately fundamental properties of glow discharges; it produces reasonably well the current–voltage characteristics (CVC) as well as specific discharge modes occurring along the CVC, and the spatial profiles of the discharge characteristics, such as the charged particle densities, electron temperature, and the electric potential and field.

2.4 PIC/MCC method for simulation of gas discharges

A correct description of such a highly non-equilibrium medium as plasma of gas discharge can be done only by kinetic theory [33]. Within this theory, the behavior of charged particles is described by the velocity distribution function (VDF) $f(\mathbf{r}, \mathbf{v}, t)$ that defines the probability distribution of particle velocity vector $\mathbf{v}$ at position $\mathbf{r}$ in space at time t. Alternative to the direct solution of the Boltzmann's equation are particle simulation methods [34]. These methods also lead to the VDF and to all macroscopic properties of plasma (the particle densities, currents, the transport coefficients and collision rates, etc), which are derived from it.

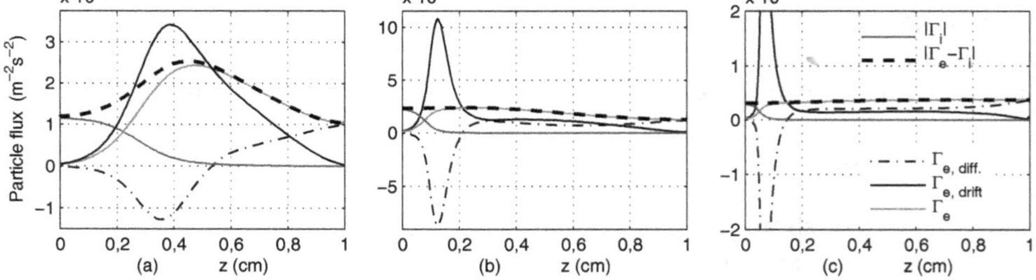

Figure 2.15. Axial profiles of axial components of ion flux, $\mathbf{\Gamma}_i$, and electron drift, diffusion, and total fluxes, $\mathbf{\Gamma}_{e,\mathrm{drift}}$, $\mathbf{\Gamma}_{e,\mathrm{diff}}$, and $\mathbf{\Gamma}_e$, and difference $\mathbf{\Gamma}_e - \mathbf{\Gamma}_i$. Panels (a), (b), and (c) correspond to the regimes 1, 4, and 7 in figure 2.10. Adapted from [46] with permission of AIP Publishing.

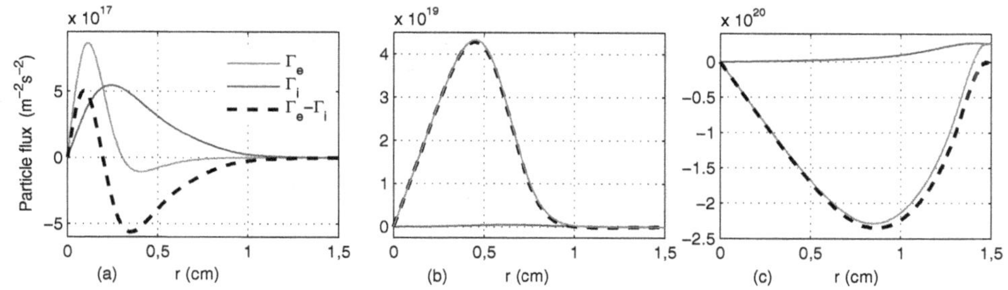

Figure 2.16. Radial profiles of radial components of ion and electron fluxes Γ_i and Γ_e and their difference $\Gamma_e - \Gamma_i$. Panels (a), (b), and (c) correspond to parameter regimes 1, 4, and 7 in figure 2.10. Adapted from [46] with permission of AIP Publishing.

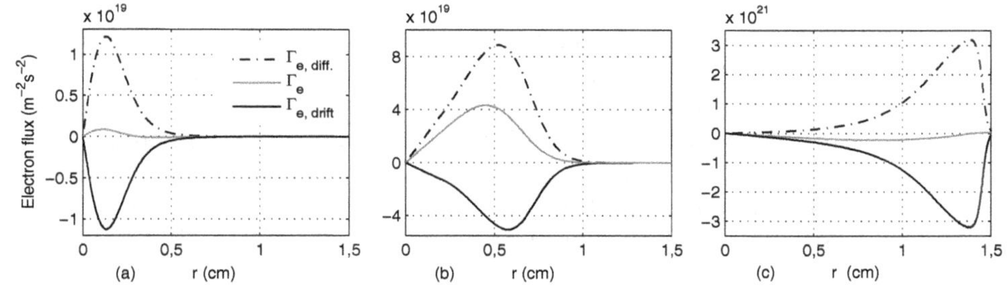

Figure 2.17. Radial profiles of radial components of electron drift, diffusion, and total fluxes, $\Gamma_{e,\text{drift}}$, $\Gamma_{e,\text{diff}}$, and Γ_e. Panels (a), (b), and (c) correspond to parameter regimes 1, 4, and 7 in figure 2.10. Adapted from [46] with permission of AIP Publishing.

In this section, the PIC (particle-in-cell) technique is described and illustrated. This is a method belonging to the class of *particle-mesh* approaches. The two basic ideas that it utilizes consist of

- using computational grid that eliminates the need to account for the pairwise interaction of all individual particles, and
- concept of *superparticles*, representing a large number of real particles, that reduces the number of particles in the simulation to a tractable scale.

The PIC is based on the simultaneous solutions of the Lorentz equations of motion for particles,

$$\dot{\mathbf{r}} = \mathbf{v}, \tag{2.111}$$

$$m\dot{\mathbf{v}} = q(\mathbf{E} + \mathbf{v} \times \mathbf{B}) \tag{2.112}$$

coupled with the Maxwell equations for electric and magnetic fields **E** and **B**. Here, q and m are the charge and the mass of the particle, respectively. In the following we restrict the analysis to electrostatic simulation, assuming $\mathbf{B} = 0$.

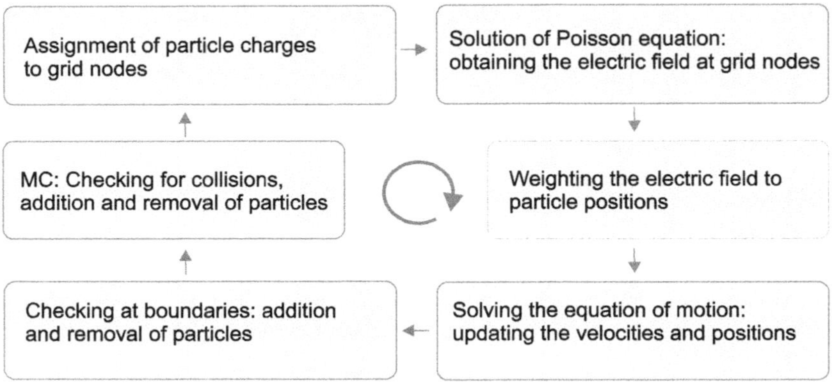

Figure 2.18. The PIC/MCC flow chart for a bounded, collisional plasma.

In PIC simulations of gas discharges, the collisions between the particles are usually taken into account by the Monte Carlo (MC) technique [35]. Here we consider only collisions of the charged particles with the atoms/molecules of the background gas, and neglect electron–electron collisions. The simulation scheme combining PIC and MC techniques is known as the Particle in cell/Monte Carlo collision (PIC/MCC) method [36]. In this section, PIC/MCC is illustrated by its application to simulation of plasma of radio frequency (RF) discharge.

2.4.1 PIC/MCC simulation method

As opposed to fluid methods, PIC/MCC method considers motion and interactions of individual charged particles (more precisely, *superparticles*) in the system. Each superparticle consist of specific number of real particles of the same type, and this number is called a *weighting* (W) of the superparticle. For bounded and collisional plasmas, PIC/MCC simulation cycle for spatially 1D system in absence of the magnetic field is summarized as follows (see figure 2.18).

(i) *Initialization.* Charged particles (superparticles) are distributed randomly within the computational domain. Uniform distribution is usually used in this step. Components v_x, v_y, and v_z of the particle velocity are specified as

$$v_s = \sqrt{-\ln(R_1) \times 2k_BT/m} \times \sin(2\pi R_2), \tag{2.113}$$

where the subscript s stands for either x, y, or z, and R_1 and R_2 are random numbers uniformly distributed in [0, 1]. This corresponds to Maxwellian velocity distribution of particle with the temperature T and mass m.

(ii) *Calculation of charged particle densities at grid points.* With the positions of superparticles known, by using one of the weighting methods, the particle densities at points of the computational grid are evaluated. Consider uniform grid with step size Δx and assume that N superparticles of type p are located between two neighboring grid points x_k and x_{k+1}. Then, by the first-order weighting [36], densities n_{p_k} and $n_{p_{k+1}}$ of particles of this type at these points are calculated from

$$n_{p_k} = W_p \sum_{j=1}^{N} \frac{\Delta x_{j,k+1}}{\Delta x \, \Delta V}, \quad n_{p_{k+1}} = W_p \sum_{j=1}^{N} \frac{\Delta x_{j,k}}{\Delta x \, \Delta V}, \tag{2.114}$$

where W_p is a weighting of particles of type p, $\Delta x_{j,k}$ and $\Delta x_{j,k+1}$ are distances from the jth particle to the kth and $(k+1)$st grid nodes, $\Delta V = \Delta x \Delta y \Delta z$ is the volume element, where Δy and Δz are defined equal one unit of length in the case where only the x space dimension is considered.

(iii) *Solution of the Poisson's equation.* For a system with M types of charged plasma components, electrostatic potential φ is calculated from the Poisson's equation

$$-\varepsilon_0 \frac{\partial^2 \varphi}{\partial x^2} = \sum_{p=1}^{M} n_p q_p, \tag{2.115}$$

where ε_0 is the permittivity of free space, q_p is the charge of particles of type p.

(iv) *Calculation of the electric field.* Once the electric potential profile φ has been computed, strength of the electric field E at the grid points is obtained from the numerical differentiation of the potential

$$E = -\frac{\partial \varphi}{\partial x}. \tag{2.116}$$

(v) *Interpolation of the electric field to positions of charged particles.* Lorentz force qE acting on the charged particles is determined from interpolation of the electric field E, which is defined on the nodes of the computational grid, to the charged particles positions.

(vi) *Solution of the Lorentz force equations.* Equations

$$\frac{dx_j}{dt} = v_j, \tag{2.117}$$

$$\frac{dv_j}{dt} = \frac{q_j}{m_j} E_j \tag{2.118}$$

are solved to determine positions x_j and velocity components v_j of the charged particles. Here, q_j, m_j, x_j, v_j, and E_j denote the charge, mass, position and velocity of the jth particle, and the electric field acting on it. By the leapfrog method [37], with Δt denoting the time step and index k denoting the time level, these equations are implemented numerically as follows:

$$v_j^{(k+1/2)} = v_j^{(k-1/2)} + \frac{q_j}{m_j} E\left(x_j^{(k)}\right)\Delta t,$$

$$x_j^{(k+1)} = x_j^{(k)} + v_j^{(k+1/2)}\Delta t.$$

(vii) *Effect of collisions and boundary conditions.* Finally, the Monte Carlo collision (MCC) method is applied to account for the effect of particle collisions and boundary processes. Particles arriving at the boundaries are identified. They can be either absorbed and neutralized, reflected or cause to emit secondary electrons from the boundaries. Then, the algorithm advances to the next time level, the process returns back to the step (ii), and the cycle is repeated. The details of implementation of the MCC procedure are given below.

It should be noted that, for example, under typical RF discharge conditions, simulations usually require a few hundred to a thousand grid points in one space dimension, and a few thousand time steps within the RF cycle. Taking into account the time consuming MC procedure carried out after each time step, even one-dimensional PIC/MCC simulations become very computationally demanding. Among the basic ways to speed up the calculations are the *parallelization of the numerical code*, the *subcycling* of ions and slow electrons, and the so-called *null collision method* [38].

MCC procedure

In gas discharge plasmas, charged particle densities are usually several orders of magnitude lower than that of the neutral background gas. Therefore, interactions between the neutral and charged particles dominate the collisional processes. Here the MCC procedure is described for the case of two-component plasma of electrons and positive ions within the neutral uniform background gas.

Considering low pressures electropositive noble gas discharges, among the reactions induced by electrons the elastic scattering, ionization, and excitation collisions are usually taken into account. Collision cross-sections of these processes define the probabilities of electrons (superelectrons) to enter into corresponding reactions over the time interval Δt according to formula

$$P_j = 1 - \exp(-n\sigma(\varepsilon_j)\, v_j \Delta t),$$

where n is the neutral background gas density, σ is the collision cross-section of the reaction, and v_j and ε_j are the velocity and kinetic energy of the jth electron. Here, the limit of the cold gas approximation is used, which assumes that the background gas atoms are at rest [35]. Within the MCC method, this probability is compared with a random number R, uniformly distributed between 0 and 1. In the case that $P_j \geqslant R$ the electron is selecting as reacting. Under realistic discharge conditions repeating this procedure to all electrons (superelectrons) is extremely time consuming. In order to speed up this procedure, the so-called *null collision method* [39], is commonly applied.

Electron collisions: modification of the kinetic energy. Let us illustrate the electron kinetic energy (and, correspondingly, the speed) modification in the following electron-induced reactions.

(a) *Elastic scattering of electrons.* In elastic collisions of electrons with neutral atoms,

$$e + A \rightarrow e + A,$$

the relative energy loss $(\Delta\varepsilon/\varepsilon)$ of electrons is described by [35]

$$\delta = 2(1 - \cos\chi)m_e/M,$$

such that the energy of scattered electrons become

$$\varepsilon_{\text{scatt.}} = (1 - \delta)\varepsilon_{\text{incid.}}.$$

Here, m_e and M are masses of electrons and target atoms,

$$\chi = \arccos(1 - 2\,R_1),$$

where R_1 defining a random number uniformly distributed in [0, 1) is the scattering angle measured in radians.

(b) *Excitation collisions of electrons.* In the excitation of atoms by electron collisions,

$$e + A \rightarrow e + A^*,$$

electrons lose the energy $\Delta\varepsilon$ corresponding to excitation from the state A to level A^*, such that the kinetic energy of the scattered electrons is obtained from

$$\varepsilon_{\text{scatt.}} = \varepsilon_{\text{incid.}} - \Delta\varepsilon.$$

(c) *Ionization collisions of electrons.* In ionization of neutral atoms by electron impact,

$$e + A \rightarrow 2\,e + A^+,$$

incident electron loses an amount $\Delta\varepsilon$ of its kinetic energy, which is spent for ionization of an atom. The remaining energy $\varepsilon_{\text{incid.}} - \Delta\varepsilon$ is shared between the scattered and emitted electrons:

$$\varepsilon_{\text{scatt.}} + \varepsilon_{\text{emitt.}} = \varepsilon_{\text{incid.}} - \Delta\varepsilon.$$

The partitioning of the energy between the scattered and emitted electrons can be assigned randomly, or, as a rough approximation, it can be divided equally between these electrons.

Electron collisions: modification of the velocity direction. Assuming that electron scattering is isotropic, the direction of scattered electron is determined by two parameters, the scattering angle $\chi = \arccos(1 - 2R_1)$ introduced above and azimuthal angle $\eta = 2\pi R_2$, where R_1 and R_2 are random numbers distributed uniformly in [0, 1). Given that the direction of electron velocity prior to collision is defined in coordinates θ and ϕ such that

$$\begin{bmatrix} v_x \\ v_y \\ v_z \end{bmatrix} = v \begin{bmatrix} \cos\theta \\ \sin\theta\cos\phi \\ \sin\theta\sin\phi \end{bmatrix}, \tag{2.119}$$

past the collision it is rotated to [35]

$$\begin{bmatrix} \tilde{v}_x \\ \tilde{v}_y \\ \tilde{v}_z \end{bmatrix} = \tilde{v} \begin{bmatrix} \cos\theta & -\sin\theta & 0 \\ \sin\theta\cos\phi & \cos\theta\cos\phi & -\sin\phi \\ \sin\theta\sin\phi & \cos\theta\sin\phi & \cos\phi \end{bmatrix} \cdot \begin{bmatrix} \cos\chi \\ \sin\chi\cos\eta \\ \sin\chi\sin\eta \end{bmatrix}. \tag{2.120}$$

Collisions of ions with neutral atoms. In collisions of ions with neutral atoms isotropic scattering

$$A + B^+ \rightarrow A + B^+$$

and charge exchange collisions, where the ion passing near the neutral atom is pulling off the electron,

$$A + B^+ \rightarrow A^+ + B,$$

are usually taken into account (for details, see e.g., [35, 40]).

The null collision method. Consider the situation where electrons can enter into K types of reactions, whose collision cross-sections are $\sigma_k(\varepsilon)$ ($k = 1, 2, \ldots, K$), and whose sum is

$$\sigma_T(\varepsilon) = \sum_{k=1}^{K} \sigma_k(\varepsilon),$$

referred to as the *total collision cross-section.* Let $\nu^* = \max\{n\sigma_T(\varepsilon)v\}$, where n is the neutral background gas density and v is the particle velocity, be the *total collision frequency* over considered ensemble of electrons. Then, the probability of an electron to enter any of these K reactions over the time interval Δt, is

$$P_{\text{null}} = 1 - \exp(-\nu^*\Delta t).$$

Therefore, given the total number N of particles in the ensemble, the maximum number of particles which will probably make collisions is NP_{null}. In order to avoid considering all particles (superparticles) individually by MCC procedure and thus speed up the calculations, according with the null collision method [39], randomly chosen fraction of NP_{null} particles is considered. From the set of K reactions, actual reaction to which every jth particle from this fraction will be probably involved over the time interval Δt, is determined from the correlation of the ratios of the collision frequencies $\nu_k(\varepsilon_j) = n\sigma_k(\varepsilon_j)v_j$ ($k = 1, 2\ldots, K$) to the total collision frequency ν^* [41]. Precisely, given the sequence

$$\frac{\nu_1(\varepsilon_j)}{\nu^*}, \quad \frac{\nu_1(\varepsilon_j) + \nu_2(\varepsilon_j)}{\nu^*}, \quad \ldots, \quad \frac{\sum_{k=1}^{K} \nu_k(\varepsilon_j)}{\nu^*},$$

the relation

$$\frac{\nu_1(\varepsilon_j) + \nu_2(\varepsilon_j) + \cdots + \nu_{l-1}(\varepsilon_j)}{\nu^*} < R_j \leqslant \frac{\nu_1(\varepsilon_j) + \nu_2(\varepsilon_j) + \cdots + \nu_l(\varepsilon_j)}{\nu^*},$$

where R_j is the random number distributed uniformly between 0 and 1, indicates the lth reaction induced by jth electron.

Constraints on the space and time steps

Within PIC/MCC method, there are restrictions imposed on the time and space steps Δt and Δx. These conditions are as follows:

- Δt must be sufficiently small to keep collision probability $1 - \exp(-\nu^* \Delta t)$ of a particle reasonably small to minimize the probability of more than one collision over the time interval Δt;
- Δt must satisfy the Courant condition, $v_{\max}(\Delta t/\Delta x) < C$, where parameter C depends on the time integration method, for example, for the explicit leapfrog method $C = 1$ [37];
- Δt must resolve electron oscillations, characterized by the plasma oscillation frequency $\omega_{\mathrm{p}} = \sqrt{e^2 \max(n_e)/m\varepsilon_0}$; restriction that is usually recommended is $\omega_p \Delta t \leqslant 0.2$ [35];
- the space step size must be of the order of the Debye length, $\Delta x \sim \lambda_{\mathrm{D}}$;
- in order to provide accurate, statistically relevant results, there must be sufficient number of superparticles per Debye length that implies that the plasma parameter $N_{\mathrm{D}} \gg 1$ and that imposes limitations on the weighting W of superparticles.

It appears that in simulations of plasma of gas discharges the most restrictive constraint is the Courant condition if it is not properly implemented. It should be noted that it is not reasonable to specify $v_{\max}$ in this condition as the actual maximum speed over the entire electron group. Instead, $v_{\max}$ can be estimated by using value of the electric potential drop across the electrodes. Alternatively, $v_{\max}$ can be taken equal to the thermal velocity of electrons, $v_{\mathrm{th}} = \sqrt{2k_{\mathrm{B}}T_h/m_e}$, where T_h is the effective temperature corresponding to high energetic electrons.

Exercise 11 (PIC/MCC)

1. Write Matlab/Octave code generating $N = 100$ particles, distributed randomly according to uniform distribution function in the interval of length $L = 1$ cm, whose velocities are distributed randomly according with (2.113), with mean value corresponding to the kinetic energy $\varepsilon = 1\text{eV}$. Plot figures depicting
   (a) number of particles versus position in space;

(b) profile of VDF.

2. Write Matlab/Octave code to simulate motion of N_e electrons in the constant electric field E. Assume that particles are non-interacting and distributed in the domain of length L initially as described in problem 1. Use $N_e = 100$, $L = 1$ cm, $\varepsilon = 1$ eV, and $E = 20$ V/m. Assume that particles arriving at the boundaries
   (a) leave the system;
   (b) reflect from boundaries.
3. Write Matlab/Octave code to simulate motion of $N_e = 100$ electrons and $N_i = 100$ argon ions in the constant electric field $E = 20$ V/m. Assume that particles are distributed in the domain of length L initially as described in problem 1, with velocities distributed randomly according with (2.113), with mean values corresponding to the kinetic energies $\varepsilon_e = 1$ eV and $\varepsilon_i = 0.025$ eV.
4. **(Project problem:** *Simulation of electron avalanche*) Write the MC numerical code simulating electron avalanche in the discharge gap filled with argon gas of uniform density, in the constant electric field. Take into consideration elastic and ionization collisions of electrons with argon atoms. Use the following parameters: the gas pressure $p = 1$ Torr, the gas temperature $T_g = 300$ K, the reduced electric field $E/N = 500$ Td (1Td $= 10^{-21}$Vm^2), the distance between the electrodes $d = 1$ cm. Consider the cases of (a) one single electron, and (b) one hundred electrons, emitted from the cathode with energy 1 eV. Consider 1D domain and assume that electrons arriving at boundaries are absorbed.
5. **(Project problem:** *Simulation of electron swarm*) Write the MC numerical code simulating electron avalanche in the discharge gap under conditions in problem 4 but with 0.5×10^6 electrons emitted from the cathode. Consider a spatially 1D domain of length $d = 1$ cm. Find the profiles of the electron mean energy, drift velocity, and the Townsend ionization coefficient (see [35]).
6. **(Project problem:** *Two stream instability*) Write Matlab/Octave code to simulate motion of particles in the two counter-streaming beams of electrons. Assume that initially particles are distributed in spaces as described in problem 1, whose velocity is distributed according with

$$f(x, v) = \frac{n_0}{2}\left\{\frac{1}{\sqrt{2\pi}v_{th}}e^{-(v-v_b)^2/2v_{th}^2} + \frac{1}{\sqrt{2\pi}v_{th}}e^{-(v+v_b)^2/2v_{th}^2}\right\},$$

where $v_b = 3$ is the mean velocity, $v_{th} = 1$ is the thermal velocity, and n_0 is the number density. Let the number of particles $N = 20000$, the domain of length $L = 100$, number of grid points is 1000, and the time step $\Delta t = 0.1$. All the parameters are dimensionless. Plot the graphs of speed v of particles versus coordinate x at $t= 0$, 5, 10, and 20.

2.4.2 PIC/MCC simulation of capacitively coupled RF discharge in argon

Consider a capacitively coupled RF discharge, where the left electrode is grounded, $\varphi|_{x=0} = 0$, while the right electrode is driven by a sinusoidal source, $\varphi|_{x=L} = V_d \cos(2\pi f t)$, where the voltage amplitude $V_d = 250$ V and frequency $f = 13.56$ MHz, t is the simulation time. The electrode gap is $L = 2.5$ cm. The background gas is argon at temperature $T_g = 350$ K and pressure $p = 10$ Pa. The secondary electron emission coefficient is $\gamma = 0.1$. The computational grid is uniform with 600 points. Simulations are done using six different superparticle weightings. The details of the simulations are given in table 2.4.

Results presented in figures 2.19–2.21 and table 2.4 correspond to the simulation time $t \approx 7.37 \times 10^{-5}$ s, where the system performs steady state oscillations. This time interval covers about 1100 simulation cycles. As can be seen from figure 2.19, decrease in the weighting W (which means increase in the number of superparticles in the system) leads to increase in the charged particle densities, which profiles tend to converge. The *saturation* occurs at about 1200 superelectrons per grid cell (see table 2.4). Note the decay in the numerical noise accompanying this process. The parameter N_D in table 2.4 corresponds to the maximum number of superparticles per Debye length evaluated at the discharge midplane where the concentration of the particles is highest.

Mean energy profiles of ions and electrons in the quasi-neutral region of the discharge are shown in figures 2.20 (b, c). As can be seen, inadequate weighting leads to overestimation of these parameters.

Finally, as is depicted in figure 2.21(b), the ion energy distribution function (IEDF) preserves its Maxwellian shape independent of the weighting number. However, the electron energy distribution function (EEDF) with inadequate weighting number in simulation cannot be estimated properly: its profile becomes highly distorted (figure 2.21(a)). Here the energy distribution functions are normalized according to

Table 2.4. Effect of the particle weighting (W) on the average number of superelectrons per cell, the number N_D of superelectrons per Debye length, the Debye length λ_D, and total number of superelectrons. Time $t \approx 7.37 \times 10^{-5}$ s. Parameters λ_D and N_D are calculated at the center of the discharge.

Simulation number	Weighting W ($\times 10^8$)	Average number of superelectrons per cell	Plasma parameter N_D	Debye length λ_D (mm)	Total number of superelectrons
1	41.7	26	231	1.74×10^{-1}	15 746
2	20.8	56	455	1.56×10^{-1}	33 822
3	5.2	288	1713	1.11×10^{-1}	173 054
4	3.0	566	2917	9.54×10^{-2}	339 532
5	2.0	896	4272	8.79×10^{-2}	537 500
6	1.5	1229	5635	8.42×10^{-2}	737 316

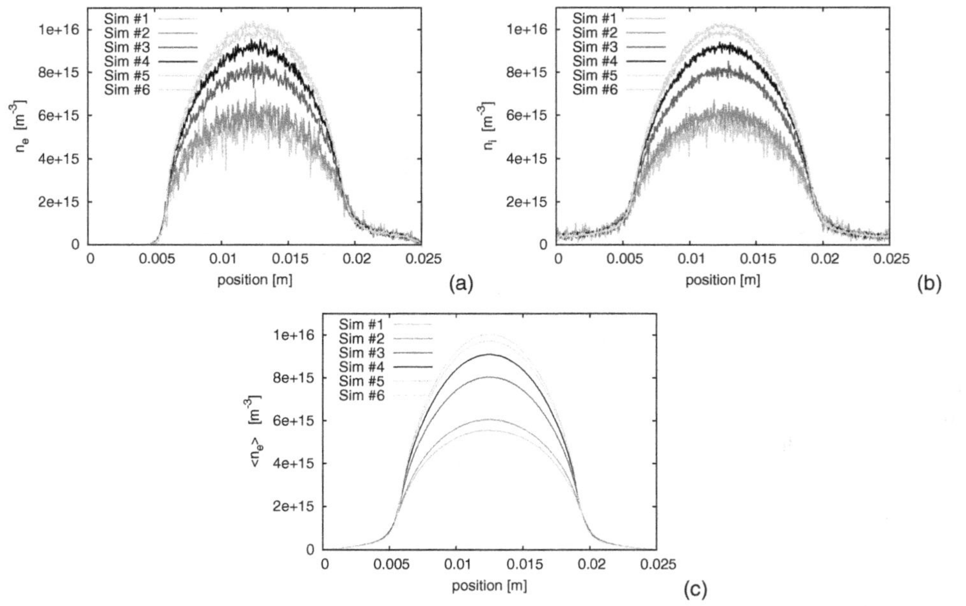

Figure 2.19. Effect of the superparticle weighting on (a) the electron density, (b) the ion density, (c) the mean electron density over one RF cycle, at time $t \approx 7.37 \times 10^{-5}$ s. Adapted from [42] with permission of Wiley.

$$\int_0^\infty \sqrt{\varepsilon}\, f(\varepsilon)\, d\varepsilon = 1. \tag{2.121}$$

In conclusion, as is demonstrated using the case of the capacitively coupled RF discharge in argon gas, the particle weighting should be taken properly into account. In fact, the discharge characteristics (such as the densities, fluxes, and average energies of charged particles, and the energy distribution functions) are highly dependent on the number of superparticles employed in the simulation. Under the considered parameter regime, the weighting independent results are obtained at about 1200 superparticles per grid cell at the discharge center, which correspond to about 5600 superparticles per Debye length in that region.

2.5 Hybrid MC–fluid modeling of gas discharges

The hybrid Monte Carlo–fluid model for a gas discharge utilizes the idea of separation of the electron population into two groups, which are the low energetic (slow) and high energetic (fast) groups [43]. Within this model, ions and slow electrons are described by the fluid approach using the drift–diffusion approximation for particle fluxes. Electric field profile is determined from the solution of the Poisson equation. Fast electrons are simulated by a suitable number (typically several hundreds) of superparticles emitted from the cathode into the discharge volume, and these are considered responsible for the ionization processes in the plasma. Collisions of fast electrons are simulated by the Monte Carlo collision

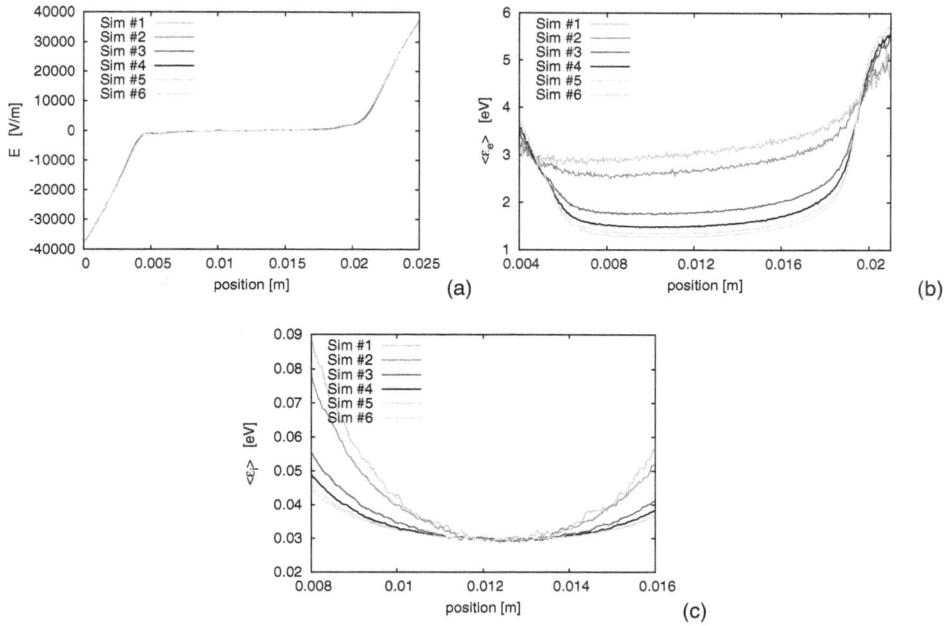

Figure 2.20. Effect of the superparticle weighting on (a) the electric field, (b) mean electron energy, (c) mean ion energy. Time $\omega t/2\pi = 0.25$ of the RF period. Adapted from [42] with permission of Wiley.

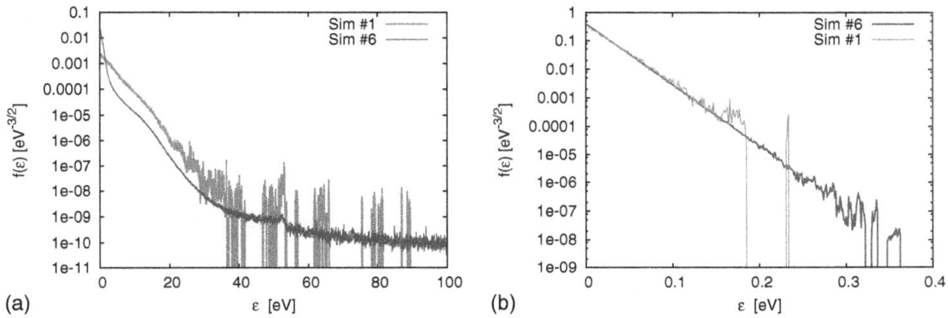

Figure 2.21. Effect of the superparticle weighting on (a) the electron energy distribution function and (b) ion energy distribution function at the center of the discharge. Time $\omega t/2\pi = 0.25$ of the RF period. Adapted from [42] with permission of Wiley.

(MCC) technique [13, 43]. Elastic, excitation and ionization collisions of fast electrons with the background neutral gas particles are usually taken into consideration, while the electron–electron and electron–ion collisions are ignored.

Fluid part of the model consists of continuity equations for slow electrons and ions [35],

$$\frac{\partial n_{se}}{\partial t} + \nabla \cdot \mathbf{\Gamma}_{se} = S_{se}, \tag{2.122}$$

$$\frac{\partial n_i}{\partial t} + \nabla \cdot \mathbf{\Gamma}_i = S_i, \tag{2.123}$$

coupled with the Poisson equation for the electrostatic field,

$$-\nabla^2 \varphi = \frac{e}{\varepsilon_0}(n_i - n_e). \tag{2.124}$$

Here, n and $\mathbf{\Gamma}$ denote the particle number and flux densities, correspondingly, and φ the electric potential. Subscripts se, fe, and i refer to slow electrons, fast electrons, and ions, respectively. The total electron density is $n_e = n_{se} + n_{fe}$. The source terms S_{se} and $S_i = S_{se} + S_{fe}$ in the right-hand sides of equations (2.122) and (2.123) describe the creation rates of slow electrons and ions, which are determined from the MCC simulation of fast electrons. The flux densities are expressed in the drift–diffusion form,

$$\mathbf{\Gamma} = -D\, \nabla n \pm \mu\, \mathbf{E}\, n, \tag{2.125}$$

where D and μ denote the diffusion and mobility coefficients, and $\mathbf{E} = -\nabla\varphi$ the electric field vector.

Calculations are carried out in the 1D domain and also in the 2D rectangular (x, y) domain with a uniform grid in each direction. The cathode is located at $x = 0$ and the anode at $x = L_x$.

Boundary conditions are defined in the following way. The electric potential is $\varphi = V_a$ on the anode, and it is grounded $\varphi = 0$ on the cathode. The dielectric side walls are assumed to be perfect absorbers. The electric field on these walls is obtained from the Gauss law, $\hat{\mathbf{n}} \cdot \mathbf{E} = -\sigma/\varepsilon_0$, where the surface charge density σ is determined from the equation

$$\frac{\partial \sigma}{\partial t} = e\,(\mathbf{\Gamma}_i - \mathbf{\Gamma}_e) \cdot \hat{\mathbf{n}}, \tag{2.126}$$

where $\hat{\mathbf{n}}$ is the normal unit vector directed outward from the wall surface.

Boundary conditions for the slow electrons and ions on the side walls and the electrodes are defined by the equations [23]

$$\hat{\mathbf{n}} \cdot \mathbf{\Gamma}_e = \frac{1}{4} n_e v_e - \alpha_e \mu_e n_e (\hat{\mathbf{n}} \cdot \mathbf{E}) - \beta_e \gamma |\hat{\mathbf{n}} \cdot \mathbf{\Gamma}_i|, \tag{2.127}$$

$$\hat{\mathbf{n}} \cdot \mathbf{\Gamma}_i = \frac{1}{4} n_i v_i + \alpha_i \mu_i n_i (\hat{\mathbf{n}} \cdot \mathbf{E}). \tag{2.128}$$

In these equations, $v_{i,e} = \sqrt{8\pi T_{i,e}/e m_{i,e}}$ denote the electron and ion thermal speeds. The last term in equation (2.127) describes the secondary electron emission from the walls due to ion impact, with the emission coefficient γ. Parameter β_e is defined to be equal to 1 on the cathode and 0 otherwise. Parameters $\alpha_{i,e}$ are responsible for regulation of the drift fluxes at the boundaries: $\alpha_{i,e} = 1$ if the flux is directed to the wall and 0 otherwise.

The fast electron fraction is treated as particles. By the MCC method, these particles are tracked in the discharge volume in the course of their motion in the electric field and making random elastic and inelastic collisions with the background neutral gas atoms. The position $\mathbf{r}_j$ and velocity $\mathbf{v}_j$ of every jth particle are obtained from the solution of equations (see section 2.4)

$$\frac{d\mathbf{r}_j}{dt} = \mathbf{v}_j, \tag{2.129}$$

$$\frac{d\mathbf{v}_j}{dt} = -\frac{e}{m_e}\mathbf{E}_j. \tag{2.130}$$

2.5.1 Spatially 1D modeling

MCC simulation of fast electrons

The electric field profile, obtained from the fluid part of the numerical code at the given time step, is used as an input parameter in the MCC cycle. The MCC cycle, in turn, generates the spatial distributions of the ionization functions for positive ions and slow electrons, which are then utilized as source terms in the corresponding continuity equations in the next time step.

Within the MCC cycle, the position x_j and velocity v_j of the jth fast electron between the collisions are updated from the equations

$$v_j^{(k+1)} = v_j^{(k)} - \frac{e}{m_e}E_{\text{eff},\, j}^{(k)}\,\Delta t, \tag{2.131}$$

$$x_j^{(k+1)} = x_j^{(k)} + v_j^{(k)}\,\Delta t - \frac{e}{2m_e}E_{\text{eff},\, j}^{(k)}\Delta t^2, \tag{2.132}$$

where the effective field is defined as [13]

$$E_{\text{eff}} = E + \frac{1}{2}\,v\frac{\partial E}{\partial x}\Delta t. \tag{2.133}$$

In these equations, Δt denotes the time step size, k is the time level, j is the electron index, e and m_e are the electron charge and mass, respectively, and v is x component of the velocity vector.

For a sufficiently small time step Δt, collision probability P_j for the jth electron can be expressed in the form

$$P_j = 1 - e^{-n\sigma_{\text{T}}(\varepsilon_j)v_j\Delta t}, \tag{2.134}$$

where n denotes the density of the background neutral gas, v_j and ε_j are the electron velocity and energy, $\sigma_{\text{T}}(\varepsilon) = \sum_{k=1}^{K}\sigma_k(\varepsilon)$ is the *total collision cross-section*, which takes into account the elastic, excitation and ionization collisions of electrons with the background neutral gas atoms. (In fact, for collisions between fast electrons and neutral gas particles, relative velocities can be taken equal to the electron velocities

[35].) A condition for a collision to occur as well as a condition for a specific type of collision (elastic, ionization or excitation) to occur is determined by using random numbers R uniformly distributed between 0 and 1. More specifically, the relation

$$\sum_{k=1}^{l-1} \sigma_k(\varepsilon_j)/\sigma_T(\varepsilon_j) < R_j \leqslant \sum_{k=1}^{l} \sigma_k(\varepsilon_j)/\sigma_T(\varepsilon_j)$$

indicates the index l of reaction induced by the jth electron.

Fast electrons lose their energy in inelastic (ionization and excitation) collisions with neutral particles as they move in the discharge volume. An electron is classified as a fast electron if it has a non-zero probability of such a collision. This is so if the total (kinetic + potential) energy of the electron is greater than the threshold excitation energy, $E_{kin} + e(\varphi_{max} - \varphi) \geqslant E_{ex}$. An electron passes from the group of fast electrons to the group of slow electrons if its total energy decreases to a value below E_{ex}.

After all fast electrons (within the MCC cycle) have left the discharge volume or passed to the group of slow electrons, the positions of relative events (such as ionizations and transitions to the slow electron group) are processed and the numbers of ions and slow electrons, N_i and N_{se}, created in each grid cell Δx are recorded. The ionization profile per secondary electron is determined by normalizing with N_0, the total number of secondary electrons emitted from cathode surface in the MCC cycle. The true ionization function profile is obtained by multiplying by the actual number of electrons emitted from cathode surface during the previous fluid cycle, so that the ionization source functions for the ions and slow electrons obtain the form

$$S_{i,se} = j_e \frac{1}{N_0} \frac{N_{i,se}}{\Delta x}, \tag{2.135}$$

where j_e is the value of the electron current density on the cathode surface. In 1D, from the condition of secondary electron emission, $j_e = j/(1 + 1/\gamma)e$, where j is the total current density, γ the secondary electron emission coefficient, and e the electron charge [14, 15].

Numerical procedure for the fluid part of the model

Consider spatially 1D implementation of the method. Continuity equations (2.122) and (2.123) are discretized using the control volume technique (see section 2.2.4), so that the corresponding discretization equations

$$\frac{n_j^{k+1} - n_j^k}{\Delta t} \Delta x + (\Gamma_{j+1/2}^{k+1} - \Gamma_{j-1/2}^{k+1}) = S_j^k \Delta x. \tag{2.136}$$

Index j indicates grid point positions, and k the time level. The flux density Γ is approximated using the exponential scheme, which is also known as Scharfetter–Gummel method [1],

$$\Gamma_{j+1/2}^{k+1} = \mu_{j+1/2} E_{j+1/2} \frac{n_j^{k+1} - n_{j+1}^{k+1} e^{-P_{j+1/2}}}{1 - e^{-P_{j+1/2}}}, \tag{2.137}$$

where P denotes the grid Peclet number,

$$P_{j+1/2} = \frac{\mu_{j+1/2}E_{j+1/2}}{D_{j+1/2}}\Delta x. \tag{2.138}$$

Equations (2.122) and (2.123) for slow electrons and ions and the Poisson equation are solved by the tridiagonal matrix algorithm (TDMA), as described in section 2.1.3. When expressed in the form suitable for solution by TDMA, equations (2.136) obtain the form

$$a_j^k n_j^{k+1} = b_j^k n_{j-1}^{k+1} + c_j^k n_{j+1}^{k+1} + d_j^k, \tag{2.139}$$

and the Poisson equation is written as

$$2\varphi_j^k = \varphi_{j-1}^k + \varphi_{j+1}^k + \Delta x^2 \omega_j^k. \tag{2.140}$$

The stationary solution of the problem is achieved by successive iterations of the fluid and Monte Carlo cycles of the model. Since the electric field profile is taken from the previous time step, the progression in time is partially implicit and partially explicit. This imposes limitations on the time step size Δt (see also *constraints on the space and time steps* in section 2.4), which are caused by

- the Courant–Friedrichs–Lewy (CFL) condition,
- the period of the electron plasma oscillations, and
- the dielectric relaxation time,

$$\tau_{DR} = \frac{\varepsilon_0}{e\mu_e n_e + e\mu_i n_i}.$$

The last condition imposes the strongest restriction on the time step, especially in the case of relatively high plasma density. For example, for plasma density about $n = 10^{10}$ cm^{-3}, τ_{DR} is of the order 10^{-10} s.

2.5.2 Results of 1D numerical implementation

The computations are carried out for DC discharge in argon gas at pressure $p = 1$ Torr and constant background temperature $T_g = 0.025$ eV. The computational domain is restricted to the longitudinal x axis orthogonal to the electrodes, the discharge cell size is $L = 1$ cm. The mobility and diffusion coefficients for slow electrons are defined by $\mu_e = 3 \times 10^5$ cm^2 s^{-1}V^{-1} and $D_e = \mu_e T_e$, where the following values for the electron temperature T_e are tested: 0.3, 1, and 3 eV [14]. The ion temperature is assumed to be equal to the background gas temperature, $T_i = T_g$. The mobility and diffusion coefficients for ions are defined as functions of reduced electric field, E/p, as in [44, 45].

In this section, results obtained from 1D hybrid Monte Carlo–fluid simulations are compared with those from [14]. A field reversal phenomena is discussed. The results corresponding to the *simple fluid* and *extended fluid* models obtained under the same conditions are also presented. Next, in section 2.5.3, we present the 2D hybrid MC–fluid modeling approach, and compare current–voltage curves, obtained with 1D and 2D models and from the experiments.

Comparison with results obtained from fluid models

1D simulations are performed under conditions given in [14], for argon at $p = 1$ Torr, the gap size between the electrodes is $L = 1$ cm, the applied voltage is $V_a = 250$ V, $T_e = 1$ eV, and the secondary electron emission coefficient is $\gamma = 0.06$. In order to compare the results with those from [14], the boundary conditions are not as described by equations (2.127)–(2.128) but imposed in a more crude way: on the anode $n_i = n_e = 0$, on the cathode $\partial n_i/\partial x = 0$ and the electron density is defined by the relation $n_e = n_i\mu_i/\mu_e$ expressing secondary emission of electrons (2.127). Computed profiles of the electron and ion densities, the electric field, and the ionization source are shown in figure 2.22. As is evident from this figure, profiles of the particle densities (panel (a)) and the electric field (panel (b)) are in excellent agreement with the results reported in [14]. The difference in the ionization function (panel (c)), which appears in the region near the anode, can be attributed to different number of superelectrons used in the Monte Carlo cycle and also to different conditions for partition of electrons into fast and slow groups. However, as is obvious from panels (a) and (b), this has no effect on the plasma properties.

Notice that figure 2.22 also illustrates the results obtained under the same parameter regime (with $V_a = 250$ V) from the simple fluid and extended fluid models. This allows observation of the effect of different fluid modeling approaches on the discharge properties. Recall that in the simple fluid model, the ionization source is approximated by the Townsend formula (see section 2.3.1). In the case of the extended fluid model, the plasma-chemical reactions were defined exactly as in section 2.3.2 [31]. The current densities corresponding to results in figure 2.22 are 0.11 mA cm^{-2} in the case of the simple fluid model and 0.8 and 1.74 mA cm^{-2} in the case of the hybrid and extended fluid models, respectively. The first one appears to be about one-tenth of that for the hybrid and extended fluid models so that the corresponding discharge regime falls into a different region of CVC (current–voltage characteristic) curve.

Discrepancy in the discharge properties obtained from different fluid models as well as from fluid and hybrid models is basically caused by comparing essentially different discharge regimes. Indeed, for a more adequate comparison, the discharge properties must be computed not at the same applied voltage, but, instead, at the same discharge current. Figure 2.23 shows solution profiles from figure 2.22, which were obtained in the case of the same discharge current density, 0.8 mA cm^{-2}, rather than the voltage. (Here, the boundary conditions are applied as described by equations (2.127)–(2.128).) The voltage required to maintain the discharge with that current density is 255 V, 223 V, and 706 V in the case of the hybrid model, extended fluid model, and simple fluid model, respectively. Vertical lines in figures 2.22 and 2.23 separate the cathode sheath and negative glow regions of the discharge, for different models. The boundaries between these regions were determined from the condition $n_e = 0.5\, n_i$.

The discharge properties in figure 2.23 computed from fluid models, compared to those in figure 2.22, are obviously qualitatively similar, with plasma density profiles

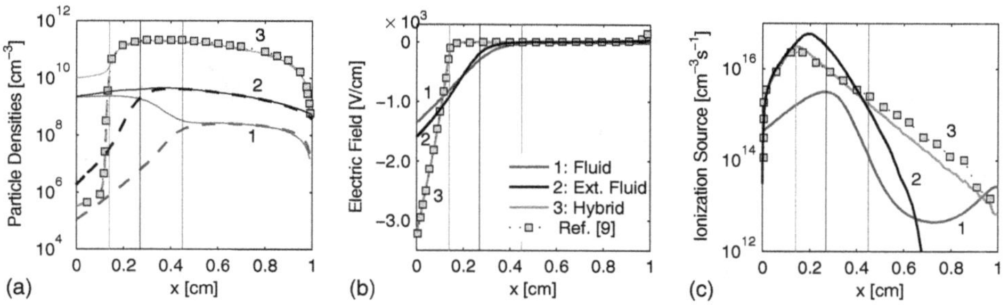

Figure 2.22. 1D model for DC discharge. (a) Electron and ion densities, (b) electric field, and (c) ionization source profiles. Squares correspond to computed results from [14]. Vertical lines indicate cathode sheath thicknesses. $p = 1$ Torr, $T_e = 1$ eV, $\gamma = 0.06$, $L = 1$ cm, $V_a = 250$ V. Adapted from [46] with permission of AIP Publishing.

close to each other, almost equal cathode sheath layers, and ionization function profiles, which are closely spaced in the cathode layer and diverge in the negative glow. However, these results differ significantly from those obtained from the hybrid model in the plasma density and the thickness of the cathode sheath layer.

It is known that electron transport in the glow discharge plasma cannot be determined as a function of the local electric field [5, 6]. This feature is also associated with non-locality of ionization in the plasma. In glow discharge models, which describe this feature correctly, in spite of weak electric field in the negative glow of the discharge, ionization events are taken into account properly in that region [5, 46]. As can be seen from figure 2.23(c), the hybrid model and partially extended fluid model predict the ionization function adequately, while the simple fluid model does not. The hybrid model takes account of the non-local character of the electron transport by means of the MCC simulation of the fast electron kinetics. Within the extended fluid model, non-locality of the electron transport is partially incorporated through the convective heat flux from the cathode layer, where the main Joule heating occurs, in the electron energy balance [20, 21].

Figure 2.23, in addition to the case with $T_e = 1$ eV shown in figure 2.22, also contains results computed from the hybrid model for $T_e = 0.3$ and 3 eV. The effect of T_e on the particle number density is reasonable: the temperature T_e is inversely proportional to the slow electron density n_e (see figure 2.23(a)). In order to maintain a discharge with current density $J = 0.8$ mA cm^{-2}, the applied voltage V_a varies as 259, 255, and 243 V for $T_e = 0.3$, 1, and 3 eV, respectively. The effect of T_e on the electric field and ionization source is weak (see figure 2.23(b, c)). The reason is that under the same current density, the charge density profiles for $T_e = 0.3$, 1, and 3 eV turn out to be very close to each other in the cathode region of the discharge, so that the resulting electric field profiles become almost identical (figure 2.23(b)). In the Monte Carlo simulation of fast electrons, which is independent of T_e, the electric field profile is used as an input parameter. Correspondingly, the ionization functions, which are determined from the Monte Carlo simulations, closely match each other. As a particular result, it is impossible to reproduce the results of the extended fluid model by a suitable choice of the slow electron temperature T_e in the hybrid model.

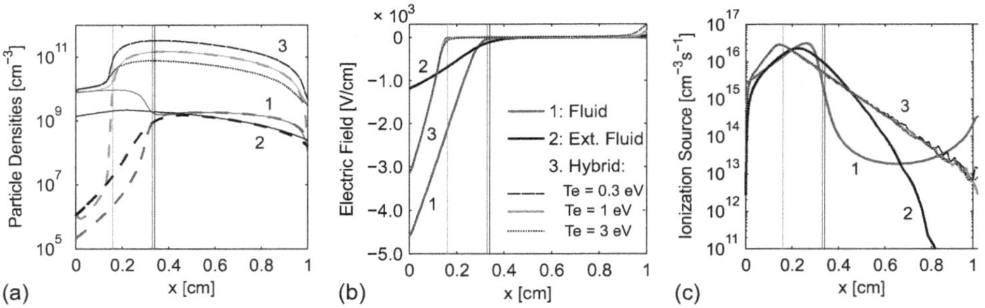

Figure 2.23. The same plasma properties as in figure 2.22 computed at the same current density. (a) Electron and ion densities, (b) electric field, (c) ionization source. Vertical lines indicate cathode sheath thicknesses. $p = 1$ Torr, $T_e = 0.3$, 1, and 3 eV, $\gamma = 0.06$, $L = 1$ cm, $J = 0.8$ mAcm^{-2}. Adapted from [46] with permission of AIP Publishing.

Electric field reversal

Reversal of the electric field in the negative glow presents an interesting effect of non-local ionization in the plasma of glow discharge. As shown in figures 2.24 (b, c), hybrid and extended fluid models reproduce a field reversal, which is not detected by the simple fluid model. Notice that the ion current reverses direction toward the anode ($-J_i < 0$) exactly at the electric field reversal point (figures 2.24 (a, b)). The uniform profile of the total current density J in figure 2.24(a) implies a correct implementation of the numerical model.

It interesting to notice that the computed field reversal point is in perfect agreement with the estimated according to [6], where the non-local ionization source was approximated by the function

$$S_i(x) = \begin{cases} \alpha_c \, \Gamma_{e0} \exp(\alpha_c x) & \text{if } x < d_c, \\ \max(S_i) \exp(-(x - d_c)/\lambda) & \text{if } x \geqslant d_c. \end{cases} \tag{2.141}$$

Here, $\alpha_c = \alpha_c(U_a/d_c)$ is the Townsend coefficient defined as function of the mean electric field over the cathode layer, U_a and d_c are the potential drop and thickness of the cathode layer, λ is the rate of decay of the ionization source (which is about the rate of decay of the fast electron flux in the negative glow), and $\max(S_i)$ is the maximum value of the ionization source (it locates about the cathode sheath boundary). Computed ionization source profile (obtained from the 1D hybrid model, figure 2.22(c)) is used to fit the parameters in equation (2.141): $\alpha_c = 1800$ cm^{-1}, $d_c = 0.15$ cm, and $\lambda = 0.10$ cm. From the equation [6]

$$x_m = d_c - \lambda \ln\left\{\frac{\lambda}{L - d_c}\left[1 - \exp\left(-\frac{L - d_c}{\lambda}\right)\right]\right\}, \tag{2.142}$$

where the field reversal point x_m is defined in terms of d_c, λ, and the discharge gap L, it follows that $x_m = 0.364$ cm. This is in excellent agreement with the position of the field reversal predicted by the hybrid model, which is 0.360 cm.

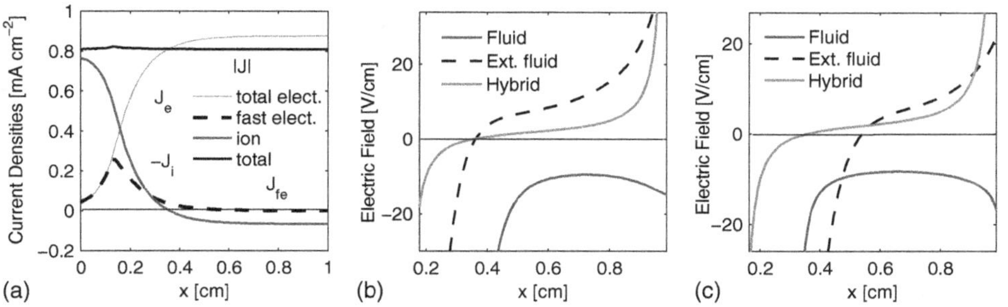

Figure 2.24. 1D hybrid model. (a) Ion current density J_i, total electron current density J_e, fast electron current density J_{fe}, and total current density J (conditions are the same as in figure 2.22), (b) closer look to the field reversal in figure 2.22(b), (c) closer look to the field reversal in figure 2.23(b). Adapted from [46] with permission of AIP Publishing.

2.5.3 Spatially 2D implementation

Drift–diffusion equations (2.122) and (2.123) are discretized by the control volume method, so that the resulting equations in 2D rectangular domain (x, y) obtain the form

$$\frac{n_{i,j}^{(k+1)} - n_{i,j}^{(k)}}{\Delta t}\Delta x \Delta y + \left(\Gamma_{xi,j+1/2}^{(k+1)} - \Gamma_{xi,j-1/2}^{(k+1)}\right)\Delta y + \left(\Gamma_{y_{i+1/2,j}}^{(k+1)} - \Gamma_{y_{i-1/2,j}}^{(k+1)}\right)\Delta x = S_{i,j}^{(k)}\Delta x \Delta y. \quad (2.143)$$

Indices i and j indicate the grid points in y and x directions, k is the time level. The components of the flux density $\mathbf{\Gamma}$ are approximated by the exponential scheme (also known as Scharfetter–Gummel method) [1],

$$\Gamma_{xi,j+1/2}^{(k+1)} = \mu_{i,j+1/2}\, E_{xi,j+1/2}\frac{n_{i,j}^{(k+1)} - n_{i,j+1}^{(k+1)}\, e^{-P_{i,j+1/2}}}{1 - e^{-P_{i,j+1/2}}}, \quad (2.144)$$

$$\Gamma_{y_{i+1/2,j}}^{(k+1)} = \mu_{i+1/2,j}\, E_{y_{i+1/2,j}}\frac{n_{i,j}^{(k+1)} - n_{i+1,j}^{(k+1)}\, e^{-P_{i+1/2,j}}}{1 - e^{-P_{i+1/2,j}}}. \quad (2.145)$$

Here, P stands for the grid Peclet number,

$$P_{i,j+1/2} = \frac{\mu_{i,j+1/2}E_{xi,j+1/2}}{D_{i,j+1/2}}\Delta x, \quad (2.146)$$

$$P_{i+1/2,j} = \frac{\mu_{i+1/2,j}E_{y_{i+1/2,j}}}{D_{i+1/2,j}}\Delta y. \quad (2.147)$$

Equations (2.143) for ions and slow electrons are solved by the TDMA line by line in the longitudinal x-direction (method of lines [1]). When written in a form suitable for solution by TDMA, these equations become

$$a_{i,j}^{(k)} n_{i,j}^{(k+1)} = b_{i,j}^{(k)} n_{i,j-1}^{(k+1)} + c_{i,j}^{(k)} n_{i,j+1}^{(k+1)} + \tilde{S}_{i,j}^{(k)}. \tag{2.148}$$

The right-hand side $\tilde{S}_{i,j}^{(k)}$ contains terms from neighboring grid points from the transverse y-direction,

$$\tilde{S}_{i,j}^{(k)} = d_{i,j}^{(k)} n_{i-1,j}^{(k)} + e_{i,j}^{(k)} n_{i+1,j}^{(k)} + S_{i,j}^{(k)}. \tag{2.149}$$

The discretized equations derived from the Poisson equation are written in the form

$$A_j^{(k)} \Phi_j^{(k)} = B_j^{(k)} \Phi_{j-1}^{(k)} + C_j^{(k)} \Phi_{j+1}^{(k)} + \Omega_j^{(k)}, \tag{2.150}$$

and solved by the block-tridiagonal matrix algorithm (see [4], p. 56). Here A, B and C are $N_y \times N_y$ matrices, Φ and Ω are $N_y \times 1$ column matrices.

The stationary solution of the problem is achieved by successive iterations of the fluid and Monte Carlo parts of the model. MCC procedure is implemented in a way similar to that of the 1D model. The MCC cycle uses as an input parameter the electric field profile obtained from the fluid part of the code in the previous time level, and it generates spatial distributions of the ionization functions for positive ions and slow electrons, which are then utilized as source terms for the respective continuity equations on the next time level.

In the 2D domain, within the MCC cycle, the positions and velocities $\mathbf{r}_l$ and $\mathbf{v}_l$ of fast electrons between the collisions are updated from the equations

$$\mathbf{v}_l^{(k+1)} = \mathbf{v}_l^{(k)} - \frac{e}{m_e} \mathbf{E}_{\mathrm{eff},\, l}^{(k)} \Delta t, \tag{2.151}$$

$$\mathbf{r}_l^{(k+1)} = \mathbf{r}_l^{(k)} + \mathbf{v}_l^{(k)} \Delta t - \frac{e}{2m_e} \mathbf{E}_{\mathrm{eff},\, l}^{(k)} \Delta t^2, \tag{2.152}$$

where the components of the effective electric field $\mathbf{E}_{\mathrm{eff}}$ are defined in the form [13]

$$E_{\mathrm{eff}\,x} = E_x + 0.5\, v_x \frac{\partial E_x}{\partial x} \Delta t, \tag{2.153}$$

$$E_{\mathrm{eff}\,y} = E_y + 0.5\, v_y \frac{\partial E_y}{\partial y} \Delta t. \tag{2.154}$$

Here, k stands for the time level and l for the electron index. Δt is the time step size, e the electron charge, m_e the electron mass, v_x and v_y are x and y components of the velocity vector. Restrictions on the step size Δt are similar to those in the 1D simulation.

Collisional processes induced by fast electrons are processed similar to that of the 1D simulation. Once all fast electrons within the MCC cycle have escaped from the discharge volume or transferred to the slow electron group, positions of relevant events (such as ionizations and transitions to the slow electron group) are registered, and numbers of ions and slow electrons, N_i and N_{se}, created within each grid cell

$\Delta V = \Delta x \Delta y$ are recorded. A profile per one secondary electron is obtained by normalizing with N_0, a total number of secondary electrons emitted from the cathode surface in the MCC cycle. The actual ionization source profile is derived by multiplying with the actual number of electrons, emitted from the cathode surface in the previous fluid cycle, so that the source functions for ions and slow electrons obtain the form

$$S_{i,se} = j_e \frac{1}{N_0} \frac{N_{i,se}}{\Delta V}. \tag{2.155}$$

In this equation j_e stands for the electron current density on the cathode. As distinct from 1D case, j_e is the function of the transversal coordinate y. It is determined from the conditions for secondary electron emission, using current density distribution on the cathode, calculated in the fluid cycle in the previous time step.

Current–voltage curves

The CVCs present important integral characteristics of the glow discharges. Computed CVCs obtained from 1D and 2D hybrid codes with constant secondary electron emission coefficient $\gamma = 0.06$ are demonstrated in figure 2.25(a). Due to charged particles escape from the side walls, voltages in the case of 2D calculations are slightly higher than those obtained from 1D [25]. Figure 2.25(a) includes also CVC computed from the 1D hybrid model from [15], which is in good agreement with present 1D result.

In the glow discharge models, secondary electron emission coefficient γ is one of the basic sources of uncertainty. It is used as a fitting parameter in some of the discharge models to match the computed data with the measured data [15]. In order to demonstrate the effect of this coefficient on the modeling results, figure 2.25(a) includes CVC computed from the hybrid model with emission coefficient dependent

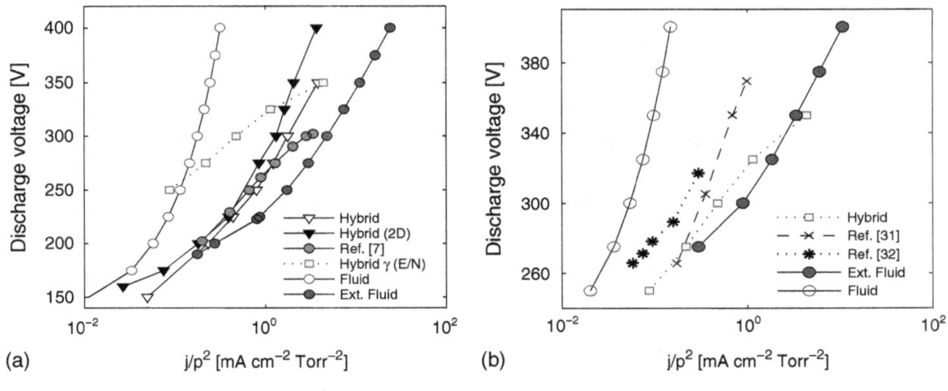

Figure 2.25. (a) CVC curves obtained from hybrid (1D and 2D) and fluid (1D) models, emission coefficients $\gamma = 0.06$ and $\gamma = 0.01(E/N)^{0.6}$, (b) comparison of CVC's computed from 1D hybrid and fluid models with $\gamma = 0.01(E/N)^{0.6}$ and measured CVCs from [47, 48]. Notice that CVC with electric field dependent γ obtained from 1D hybrid model (yellow squares) appears in both panels (a) and (b). Adapted from [46] with permission of AIP Publishing.

on the reduced electric field, $\gamma = 0.01(E/N)^{0.6}$, as suggested in [49]. In this case CVC has a slope essentially different from those obtained with constant $\gamma = 0.06$. This, however, turns out to be much closer to the measured CVCs from [47, 48] as depicted in figure 2.25(b).

Notice that figure 2.25 also includes CVCs computed from the fluid models. These figure illustrates evident failure of simple fluid models to predict adequately integral characteristics of the discharge. As can be seen, CVC curves computed from the simple fluid model are significantly further apart from those obtained from the measurements as well as from the hybrid and extended fluid models. Compared to CVC curves computed from the extended fluid model, CVCs from the hybrid model locate relatively closer to those from the measurements. However, the comparison does not reveal significant advantages of the hybrid method over the extended fluid model as reported in [14]. Accuracy of the integral characteristics for both hybrid and extended fluid models is a matter of fitting parameter γ as shown in figure 2.25. Concerning the local characteristics, hybrid models are obviously able to describe the cathode region more accurately [14].

References

[1] Patankar S 1980 *Numerical Heat Transfer and Fluid Flow* (New York: McGraw-Hill)

[2] MATLAB 2018 version 9.4.0.813654 (R2018a) (Natick, MA: The MathWorks Inc.)

[3] Fletcher C A J 1988 *Computational Techniques for Fluid Dynamics* vol 1 (Berlin: Springer)

[4] Jardin S 2010 *Computational Methods in Plasma Physics* (Boca Raton, FL: CRC Press)

[5] Kolobov V I and Tsendin L D 1992 Analytic model of the cathode region of a short glow discharge in light gases *Phys. Rev.* A **46** 7837

[6] Kudryavtsev A A, Morin A V and Tsendin L D 2008 Role of nonlocal ionization in formation of the short glow discharge *Tech. Phys.* **53** 1029

[7] Winkler R, Arndt S, Loffhagen D, Sigeneger F and Uhrlandt D 2004 Progress of the electron kinetics in spatial and spatiotemporal plasma structures Contrib *Plasma Phys.* **44** 437–49

[8] Kolobov V I and Arslanbekov R R 2006 Simulation of electron kinetics in gas discharges *IEEE Trans. Plasma Sci.* **34** 895–909

[9] Robson R E, White R D and Petrović Z L 2005 Colloquium: Physically based fluid modeling of collisionally dominated low-temperature plasmas *Rev. Mod. Phys.* **77** 1303

[10] White R D, Robson R E, Dujko S, Nicoletopoulos P and Li B 2009 Recent advances in the application of Boltzmann equation and fluid equation methods to charged particle transport in non-equilibrium plasmas *J. Phys. D: Appl. Phys.* **42** 194001

[11] Fiala A, Pitchford L C and Boeuf J P 1994 Two-dimensional, hybrid model of low-pressure glow discharges *Phys. Rev.* E **49** 5607

[12] Kushner M J 2009 Hybrid modelling of low temperature plasmas for fundamental investigations and equipment design *J. Phys. D: Appl. Phys.* **42** 194013

[13] Surendra M, Graves D B and Jellum G M 1990 Self-consistent model of a direct-current glow discharge: treatment of fast electrons *Phys. Rev.* A **41** 1112

[14] Derzsi A, Hartmann P, Korolov I, Karacsony J, Bánó G and Donkó Z 2009 On the accuracy and limitations of fluid models of the cathode region of dc glow discharges *J. Phys. D: Appl. Phys.* **42** 225204

[15] Donkó Z, Hartmann P and Kutasi K 2006 On the reliability of low-pressure dc glow discharge modelling *Plasma Sources Sci. Technol.* **15** 178
[16] Van Dijk J, Kroesen G M W and Bogaerts A 2009 Plasma modelling and numerical simulation *J. Phys. D: Appl. Phys.* **42** 190301
[17] Raizer Y P 1991 *Gas Discharge Physics* (Berlin: Springer)
[18] Boeuf J-P 1987 Numerical model of RF glow discharges *Phys. Rev.* A **36** 2782
[19] COMSOL Multiphysics® v. 5.5 (Stockholm: COMSOL AB) www.comsol.com
[20] Boeuf J P and Pitchford L C 1995 Two-dimensional model of a capacitively coupled RF discharge and comparisons with experiments in the gaseous electronics conference reference reactor *Phys. Rev.* E **51** 1376
[21] Arslanbekov R R and Kolobov V I 2003 Two-dimensional simulations of the transition from Townsend to glow discharge and subnormal oscillations *J. Phys. D: Appl. Phys.* **36** 2986
[22] Grubert G K, Becker M M and Loffhagen D 2009 Why the local-mean-energy approximation should be used in hydrodynamic plasma descriptions instead of the local-field approximation *Phys. Rev.* E **80** 036405
[23] Hagelaar G J M and Pitchford L C 2005 Solving the Boltzmann equation to obtain electron transport coefficients and rate coefficients for fluid models *Plasma Sources Sci. Technol.* **14** 722
[24] Sakiyama Y, Graves D B and Stoffels E 2008 Influence of electrical properties of treated surface on RF-excited plasma needle at atmospheric pressure *J. Phys. D: Appl. Phys.* **41** 095204
[25] Bogdanov E A, Adams S F, Demidov V I, Kudryavtsev A A and Williamson J M 2010 Influence of the transverse dimension on the structure and properties of dc glow discharges *Phys. Plasmas* **17** 103502
[26] Bogdanov E A, Kapustin K D, Kudryavtsev A A and Chirtsov A S 2010 Different approaches to fluid simulation of the longitudinal structure of the atmospheric-pressure microdischarge in helium *Tech. Phys.* **55** 1430–42
[27] Becker M M, Loffhagen D and Schmidt W 2009 A stabilized finite element method for modeling of gas discharges *Comput. Phys. Commun.* **180** 1230–41
[28] COMSOL Multiphysics® v. 5.5 2019 Plasma Module User's Guide (Stockholm: COMSOL AB)
[29] Hagelaar G J M, De Hoog F J and Kroesen G M W 2000 Boundary conditions in fluid models of gas discharges *Phys. Rev.* E **62** 1452
[30] Gorin V V, Kudryavtsev A A, Yao J, Yuan C and Zhou Z 2020 Boundary conditions for drift–diffusion equations in gas-discharge plasmas *Phys. Plasmas* **27** 013505
[31] Rafatov I, Bogdanov E A and Kudryavtsev A A 2012 On the accuracy and reliability of different fluid models of the direct current glow discharge *Phys. Plasmas* **19** 033502
[32] Phelps A V 2001 Abnormal glow discharges in Ar: experiments and models *Plasma Sources Sci. Technol.* **10** 329
[33] Tsendin L D 2010 Nonlocal electron kinetics in gas-discharge plasma *Phys.-Usp.* **53** 133
[34] Birdsall C K and Langdon A B 2018 *Plasma Physics via Computer Simulation* (Boca Raton, FL: CRC Press)
[35] Donkó Z 2011 Particle simulation methods for studies of low-pressure plasma sources *Plasma Sources Sci. Technol.* **20** 024001
[36] Birdsall C K 1991 Particle-in-cell charged-particle simulations, plus Monte Carlo collisions with neutral atoms, PIC-MCC *IEEE Trans. Plasma Sci.* **19** 65–85

[37] Vahedi V, DiPeso G, Birdsall C K, Lieberman M A and Rognlien T D 1993 Capacitive RF discharges modelled by particle-in-cell Monte Carlo simulation. I. Analysis of numerical techniques *Plasma Sources Sci. Technol.* **2** 261

[38] Kawamura E, Birdsall C K and Vahedi V 2000 Physical and numerical methods of speeding up particle codes and paralleling as applied to RF discharges *Plasma Sources Sci. Technol.* **9** 413

[39] Verboncoeur J P 2005 Particle simulation of plasmas: review and advances *Plasma Phys. Control. Fusion* **47** A231

[40] Phelps A V 1994 The application of scattering cross sections to ion flux models in discharge sheaths *J. Appl. Phys.* **76** 747–53

[41] Vahedi V and Surendra M 1995 A Monte Carlo collision model for the particle-in-cell method: applications to argon and oxygen discharges *Comput. Phys. Commun.* **87** 179–98

[42] Erden E and Rafatov I 2014 *Contrib. Plasma Phys.* **54** 626

[43] Boeuf J-P and Pitchford L C 1991 Pseudospark discharges via computer simulation *IEEE Trans. Plasma Sci.* **19** 286–96

[44] McDaniel E W 1964 *Collision Phenomena in Ionized Gases* (New York: Wiley)

[45] Frost L S 1957 Effect of variable ionic mobility on ambipolar diffusion *Phys. Rev.* **105** 354

[46] Rafatov I, Bogdanov E A and Kudryavtsev A A 2012 Account of nonlocal ionization by fast electrons in the fluid models of a direct current glow discharge *Phys. Plasmas* **19** 093503

[47] Rózsa K, Gallagher A and Donkó Z 1995 Excitation of Ar lines in the cathode region of a dc discharge *Phys. Rev.* E **52** 913

[48] Stefanović I and Petrović Z L 1997 Volt ampere characteristics of low current dc discharges in Ar, H_2, CH_4 and SF_6 *JPN J. Appl. Phys.* **36** 4728

[49] Donkó Z 2001 Apparent secondary-electron emission coefficient and the voltage-current characteristics of argon glow discharges *Phys. Rev.* E **64** 026401

Chapter 3

Numerical analysis of non-linear dynamics and transition to chaos in a gas discharge–semiconductor system

Pattern forming systems are well-known in various fields of science, including biology, chemistry, and physics [1]. Of particular interest are the pattern formation phenomena in electronic media, mainly in non-linear gas discharge systems [2–6]. We focus on the study of spatial and temporal patterns in a *barrier* DC discharge with a large aspect ratio of the discharge cell. This is a system consisting of two planar parallel layers: a planar glow discharge layer that is coupled to a semiconductor layer, where the extension of the layers in the transverse direction is much greater than in the forward direction. This system is sandwiched between two planar electrodes to which a DC voltage is applied (figure 3.1).

Such a gas discharge–semiconductor system (GDSS) has been the subject of intensive experimental research (see, e.g., [7–20]). The investigations on the self-organization of the GDSS have been recently reviewed in [5, 21]. Most of the experiments on GDSS were conducted by the groups of Purwins (Munich, Germany) and Astrov (St Petersburg, Russia). An overview of experimental results on self-organization in GDSS was summarized in [2]. It was found that this system is capable of exhibiting a wide variety of patterns with spatial and spatio-temporal structure. It is interesting to note that these structures show striking similarities to those observed in various physical, chemical, or biological systems, such as Belousov–Zhabotinski reaction and Rayleigh–Benard convection, patterns in bacterial colonies, etc. Among these structures are homogeneous stationary and oscillating modes, patterns with a spatial and spatio-temporal structure, which are classified as hexagons [7–10], stripes [13, 14], rings, and spirals [9, 15]. Structures can also be chaotic [16, 22–24] and develop in the form of solitary waves [17, 18]. Most of these patterns were observed under the cryogenic conditions, when the system was cooled with liquid nitrogen to a temperature of T_0 = 90–100 K.

doi:10.1088/978-0-7503-2360-4ch3

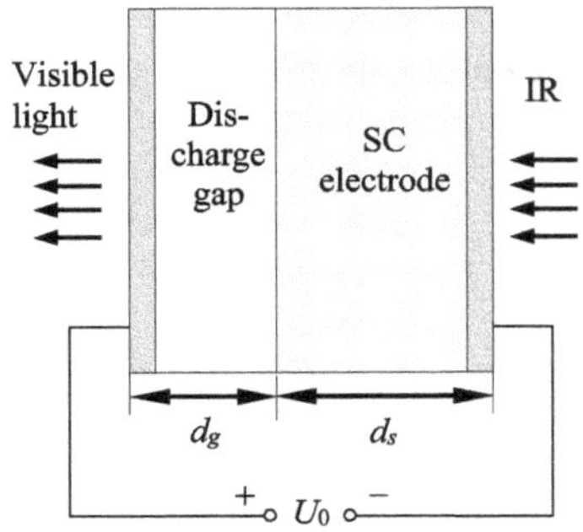

Figure 3.1. Schematic representation of the DC planar gas discharge cell.

The bifurcation diagram as well as Lorenz map showing the transition of the GDSS to chaos through a cascade of period-doubling bifurcations are discussed in section 3.2. Interestingly, the transition to chaos exhibits a perfect period-doubling cascade, strikingly similar to that in simple non-linear mappings such as the well-known logistic map. Although aperiodic oscillations and the onset of period-doubling bifurcations in the GDSS were observed before, the analysis in [25], in fact, completely and rigorously established that the GDSS falls into chaotic motion under certain parameter regimes.

One of the forms of self-organization in GDSS is typified by the presence of spatially localized solitary objects. These objects are usually bifurcated from the homogeneous stationary state of the system by the Turing destabilization mechanism. In dynamical terms, they are known as autosolitons, dissipative solitons (DSs) or spots [4, 26–28]. In response to changes in control parameters, DSs may bifurcate from stationary to traveling DSs, which can scatter from each other and generate additional DSs during this process [28–30]. Among the distinguishing features of DSs is their ability to divide. This phenomenon has been observed in the reaction–diffusion systems, where the evolution of multiplication in the pattern that 'consist of spots that grow until they reach a critical size, at which time they divide in two' [31] has been observed. The general regularities of the process of division of DSs in these systems are discussed in [27, 32].

Formation of patterns of spatially localized solitary objects in GDSS under cryogenic conditions is considered in section 3.3. These objects are generated in this system in the form of self-organized current filaments. Multiple steady-state solutions, revealed in this system, are considered in section 3.3.3. Computed results support the evidence that under the considered conditions the thermal mechanism is responsible for the instability in GDSS. This is also confirmed by the linear stability analysis of the homogeneous stationary state: when the gas heating is excluded, the instability does not form a spatial structure, it is homogeneous and growing without oscillations.

An important point is that the mechanism of the discharge instability, which leads to filamentation of the current in GDSS under cryogenic conditions, is completely different from that in discharges with a dielectric barrier (DBD). In contrast to

DBD, where the formation of patterns is caused by the field redistribution induced by the space charge, the thermal mechanism is responsible for the discharge instability and filamentation in the case of GDSS, as it was proposed in [33] and further confirmed in [34–36].

Section 3.3.5 discusses the processes of spontaneous division of current filaments (DSs) in GDSS under cryogenic conditions. Experimental studies have shown that an increase in total current in the GDSS can lead to an increase in the number of current-carrying filaments in one of two ways. The first is the pattern self-completion process [7]. This is accompanied by an increase in disturbance in the area surrounding the primary filaments, followed by the emergence of new filaments in the vicinity of the old ones. Another observed scenario is that the number of current filaments is increased by splitting of existing filaments. The process of division of DSs in the gas discharge–semiconductor system was originally observed experimentally and reported in [22, 37]. In [18], the division was observed for filaments propagating along the active area of the system (at a constant set of parameters), and for initially stationary filaments and for patterns consisting of an ensemble of DSs. In the last two cases, the spontaneous splitting of DSs was initiated by a smooth increase in the current in the system by the increase in either the applied voltage or the conductivity of the semiconductor electrode (by an increase in the intensity of IR radiation), which served as control parameters of the system.

3.1 Model

The gas discharge model used in this section is based on the diffusion–drift theory of gas discharges for two-component plasma, as it was described in section 2.3.1 (see also [23, 24, 38, 39]).

3.1.1 Governing equations

The model includes the continuity equations for charged species, namely, the electrons and positive ions with number densities n_e and n_i,

$$\frac{\partial n_e}{\partial t} + \nabla \cdot \mathbf{\Gamma}_e = S_e, \tag{3.1}$$

$$\frac{\partial n_i}{\partial t} + \nabla \cdot \mathbf{\Gamma}_i = S_i, \tag{3.2}$$

and the Poisson's equation for the electric field,

$$\nabla \cdot \mathbf{E} = \frac{e}{\varepsilon_0}(n_i - n_e), \quad \mathbf{E} = -\nabla\varphi. \tag{3.3}$$

Here, φ is the electric potential, $\mathbf{E}$ is the electric field, e is the electron charge, ε_0 is the dielectric constant. $\mathbf{\Gamma}_e$ and $\mathbf{\Gamma}_i$ are the charged particle flux densities, which consist of the drift and diffusion components,

$$\mathbf{\Gamma}_e = -n_e\mu_e\,\mathbf{E} - D_e\nabla n_e, \;\; \mathbf{\Gamma}_i = n_i\mu_i\,\mathbf{E} - D_i\nabla n_i, \tag{3.4}$$

where μ_e, μ_i and D_e, D_i are the mobility and diffusion coefficients. The current density in the discharge is

$$\mathbf{J}_g = e(\mathbf{\Gamma}_i - \mathbf{\Gamma}_e). \tag{3.5}$$

Subscripts e and i refer to electrons and ions, while subscripts s and g to semiconductor and gas discharge layers, respectively.

The source terms in the continuity equations (3.1) and (3.2) in accordance with the classical Townsend approximation are determined in the form

$$S_e = S_i = Ap\,|\mathbf{\Gamma}_e|\exp(-pB/|\mathbf{E}|), \tag{3.6}$$

where p is the pressure, A and B are constants defined by the type of the gas [40].

The semiconductor layer of thickness d_s is assumed to have a homogeneous and field-independent conductivity σ_s and dielectric constant ε_s. From the equation

$$\nabla \cdot \mathbf{J}_s = 0, \tag{3.7}$$

where the current density $\mathbf{J}_s = \sigma_s\mathbf{E}$, it follows that in the semiconductor region

$$\nabla \cdot \mathbf{E} = 0. \tag{3.8}$$

3.1.2 Boundary conditions

As shown in figure 3.2, the x-axis defines the direction parallel to the gas discharge and semiconductor layers, and the z-axis the direction normal to these layers. The anode locates at $z = 0$, the cathode at $z = d_g$, the semiconductor extends up to $z = d_g + d_s$, and the lateral boundaries are at $x = 0$ and $x = l$.

On the anode, $z = 0$, we set

$$n_i = 0, \quad \frac{\partial n_e}{\partial z} = 0. \tag{3.9}$$

On the gas discharge–semiconductor interface, $z = d_g$,

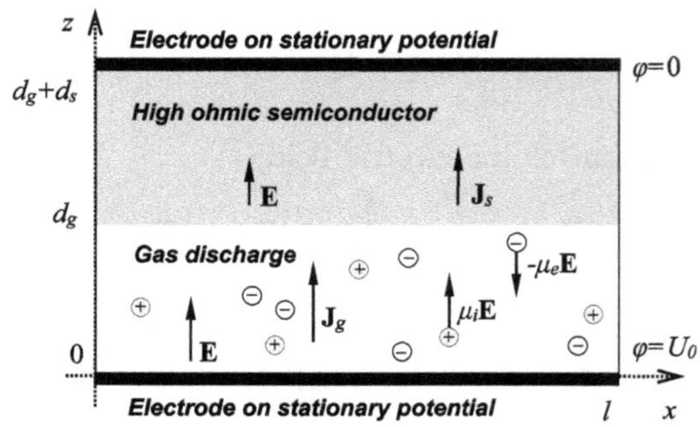

Figure 3.2. A cross-section of a discharge cell: it consist of a metal anode, a gas discharge layer, a high-ohmic semiconductor, and another metal contact.

$$\frac{\partial n_i}{\partial z} = 0, \quad \mu_e n_e = \gamma \mu_i n_i, \tag{3.10}$$

where the second condition describes the γ-process, $\mathbf{\Gamma}_e|_{z=d_g} \cdot \hat{\mathbf{n}} = -\gamma \, \mathbf{\Gamma}_i|_{z=d_g} \cdot \hat{\mathbf{n}}$, provided that the diffusion transport is neglected, γ is the secondary electron emission coefficient.

Across the interface between the gas and semiconductor layers, $z = d_g$, the electric potential is continuous,

$$\varphi\left(x,\, z = d_g^-,\, t\right) = \varphi\left(x,\, z = d_g^+,\, t\right).$$

The surface charge density is defined by the jump in the normal component of the electric field,

$$Q = \left(\varepsilon_s \varepsilon_0 \mathbf{E}|_{z=d_g^+} - \varepsilon_0 \mathbf{E}|_{z=d_g^-}\right) \cdot \hat{\mathbf{n}}, \tag{3.11}$$

and its time evolution is determined from the equation

$$\partial_t Q = \left(\mathbf{J}_g - \mathbf{J}_s\right) \cdot \hat{\mathbf{n}}, \tag{3.12}$$

where $\hat{\mathbf{n}}$ is a unit vector normal to the boundary surface and directed from the gas to the semiconductor.

Denoting applied DC voltage as U_0, we set

$$\varphi(x,\, 0,\, t) = 0, \quad \varphi\left(x,\, d_g + d_s,\, t\right) = -U_0. \tag{3.13}$$

On the lateral boundaries of the domain, $x = 0$ and $x = l$, we impose the conditions

$$\Gamma_{e,x} = \Gamma_{i,x} = \frac{\partial \varphi}{\partial x} = 0. \tag{3.14}$$

3.2 Non-linear oscillations and transition to chaos in a gas discharge–semiconductor system

In general, chaotic behavior in dynamical systems can be classified into temporal *low-dimensional* chaos associated with oscillations, and spatio-temporal chaos associated with waves [41]. The latter is also referred to as weakly developed turbulence [42]. Period multiplication, quasiperiodicity, and intermittency are usually regarded as standard pathways for a system transition to *low-dimensional* chaos [43–46]. The Lorenz map and bifurcation diagram, which demonstrate the transition of the GDSS from a periodic oscillatory regime to an irregular chaotic state through a cascade of period-doubling bifurcations, are discussed in this section.

3.2.1 Reducing of model equations and non-dimensionalization

In this section, analysis is spatially 1D, restricted to direction z orthogonal the semiconductor and discharge gap layers. In this situation the model for GDSS can be slightly simplified. Indeed, the effect of the semiconductor cathode can be incorporated into the model through equation of the form equivalent to that for the RC circuit. For a semiconductor layer of thickness d_s, with a dielectric constant ε_s and conductivity σ_s, the total current density is

$$\varepsilon_s \varepsilon_0 \frac{\partial E_s(t)}{\partial t} + J_s(t) = J(t), \tag{3.15}$$

where $J_s(t) = \sigma_s E_s(t)$. The total current $J(t)$ in the semiconductor is the same as in the gas discharge,

$$\varepsilon_0 \frac{\partial E}{\partial t} + e(\Gamma_i - \Gamma_e) = J(t). \tag{3.16}$$

With $U_s(t) = R_s J_s(t)$ and $U_s(t) = E_s(t) d_s$, we obtain from equation (3.15)

$$T_s \frac{\partial U(t)}{\partial t} = U_t - U(t) - R_s J(t), \tag{3.17}$$

where $U(t) = \int_0^{d_g} E(x, t) dx$ is the discharge voltage, $U_t = U(t) + U_s(t)$ is the total (feeding) voltage, $R_s = d_s/\sigma_s$ is the resistance, $C_s = \varepsilon_s \varepsilon_0 / d_s$ is the capacitance per area, and $T_s = C_s R_s = \varepsilon_s \varepsilon_0 / \sigma_s$ is the Maxwell relaxation time of the semiconductor.

Further, the governing equations are non-dimensionalized using the following dimensionless coordinates, time, and fields [39]:

$$\tilde{z} = \frac{z}{r_0}, \quad \tau = \frac{t}{t_0}, \quad \sigma(\tilde{z}, \tau) = \frac{n_e(z, t)}{n_0}, \quad \rho(\tilde{z}, \tau) = \frac{n_i(z, t)}{n_0},$$

$$\mathcal{E}(\tilde{z}, \tau) = \frac{E(z, t)}{E_0}, \quad \phi(\tilde{z}, \tau) = \frac{\varphi(z, t)}{E_0 r_0}.$$

The characteristic scales are $t_0 = 1/(\alpha_0 \mu_e E_0)$, $n_0 = \varepsilon_0 \alpha_0 E_0 / e$, and $r_0 = 1/\alpha_0$, where $\alpha_0 = Ap$ and $E_0 = Bp$. The dimensionless length of the discharge gap is $l = d_g / r_0$. The dimensionless parameters of the semiconductor are conductivity $\tilde{\sigma}_s = \sigma_s / (e \mu_e n_0)$ and width $l_s = d_s / r_0$, from which the resistivity is $\mathcal{R}_s = l_s / \tilde{\sigma}_s$. The applied voltage is rescaled as $\mathcal{U}_t = U_t / (E_0 r_0)$.

Then, the discharge is coupled to the semiconductor and the DC voltage source $\mathcal{U}_t$ through equation

$$\tau_s \frac{\partial \mathcal{U}(\tau)}{\partial \tau} = \mathcal{U}_t - \mathcal{U}(\tau) - \mathcal{R}_s j(\tau), \tag{3.18}$$

where $\tau_s = T_s / t_0$, $j(\tau) = \partial \tau \mathcal{E} + (\mu \rho + \sigma) \mathcal{E}$, $\mu = \mu_i / \mu_e$, and the voltage $\mathcal{U}(\tau) = \int_0^l \mathcal{E}(\tilde{z}, \tau) d\tilde{z}$ is related to the electric field $\mathcal{E}$ and potential ϕ through $\mathcal{E}(\tilde{z}, \tau) = -\partial \phi(\tilde{z}, \tau) / \partial \tilde{z}$ and $\mathcal{U}(\tau) = \phi(0, \tau) - \phi(l, \tau)$.

3.2.2 Input parameters

The choice of the input parameters is guided by the experiments in [11, 12] and specified as in [23, 38, 39]. The discharge is sustained in nitrogen at 40 mbar. Parameters $A = 12\ \text{cm}^{-1}\ \text{Torr}^{-1}$ and $B = 342\ \text{V cm}^{-1}\ \text{Torr}^{-1}$ for nitrogen [33]. In this situation, $\alpha_0 = (27.8\ \mu\text{m})^{-1}$, $E_0 = 10.3\ \text{kV cm}^{-1}$. The semiconductor layer consists of GaAs with dielectric constant $\varepsilon_s = 13.1$. The secondary emission coefficient is defined as $\gamma = 0.08$. The ion and electron mobilities are $\mu_i = 23.33\ \text{cm}^2/(\text{sV})$ and $\mu_e = 6666.6\ \text{cm}^2/(\text{sV})$. We set a dimensionless discharge gap size $L = 50$, semiconductor resistance $\mathcal{R}_s = 8.742 \times 10^6$, and a total voltage $\mathcal{U}_t$ in the range between 18 and 20 [23, 24]. These values correspond to a discharge gap 1.4 mm, semiconductor conductivity $\sigma_s = (2.6 \times 10^5\ \Omega\ \text{cm})^{-1}$, and a voltage range between 513 and 570 V.

Under these conditions, the resistance of the semiconductor layer together with the applied voltage constrains the operation to the transition mode from Townsend to glow discharge.

3.2.3 Transition from periodical to fully chaotic oscillations

Chaos and turbulence are generally considered to be undesirable features of plasma devices. For this reason, the problem of understanding and controlling the complex chaotic behavior commonly observed in gas discharge plasma is of great importance. Period-doubling bifurcations represent a typical route to (low-dimensional) temporal chaos. It has been observed in many gas discharge systems (see, e.g., [41, 47, 48]), including the gas discharge with high-ohmic semiconductor electrode [22, 49]. Occurrence of period-doubling bifurcations in GDSS was first revealed in the analysis [23]. Infinitely many period-doubling bifurcations, leading to chaos, have been detected in [24].

The bifurcation diagram constructed under parameter regime in section 3.2.2 is presented in figure 3.3 (lower panel). This diagram was generated by depicting the local maxima of the oscillation amplitude of the total discharge current $j(\tau)$ as functions of the voltage $\mathcal{U}_t$, which was varied by very small steps (of the order of 10^{-3}) from 18 to 20. (A bifurcation diagram of similar form can also be obtained by plotting the voltage $\mathcal{U}(\tau)$ versus $\mathcal{U}_t$.) For each value of the voltage $\mathcal{U}_t$, these calculations involved finding the corresponding stationary solution, and then examining the oscillations, caused by the perturbation of the system from this stationary state, far away from the initial transient phase.

The bifurcation diagram in figure 3.3 displays the classical period-doubling cascade associated with the transition to chaos according to the famous Feigenbaum scenario [50]. This diagram clearly distinguish the regions where GDSS behaves regularly and chaotically. More precisely, this diagram demonstrates the transition of the system from the regime of periodic oscillations to an irregular chaotic state with an increase in the bifurcation parameter, which in the present situation is the total voltage $\mathcal{U}_t$. In dynamical terms, during this process a new limit cycle arises from the existing one, such that the period of the new limit cycle is twice as long as that of the old one. This is illustrated in the top rows of figure 3.3, which

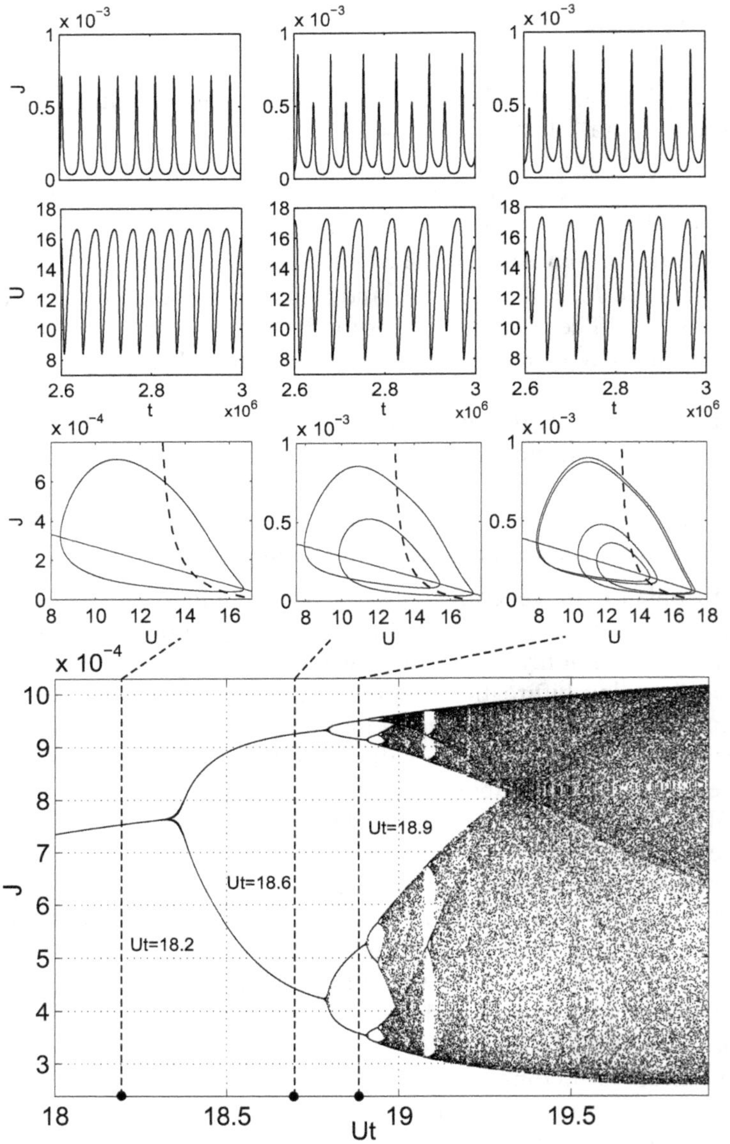

Figure 3.3. The bifurcation diagram: the discharge current j versus the total voltage $\mathcal{U}_t$ (lower panel). The top rows show the time oscillations of current j and voltage $\mathcal{U}$, and phase space trajectories of the oscillations in the plane of j and $\mathcal{U}$ for the total voltages $\mathcal{U}_t = 18.2$, 18.6, and 18.9. Two additional lines are the load line $\mathcal{U} = \mathcal{U}_t - \mathcal{R}_s j$ and the current–voltage characteristic $\mathcal{U} = \mathcal{U}(j)$ of the gas discharge. The intersection of the load line and current–voltage characteristic indicates the stationary solution of the system. Adapted from [24].

depict current $j(\tau)$ and voltage $\mathcal{U}(\tau)$ oscillations as functions of time τ, as well as phase space trajectories of the oscillations in the plane spanned by the current j and the voltage $\mathcal{U}$ for several values of the total applied voltage: $\mathcal{U}_t = 18.2$, 18.6, and 18.9. The initial transients in these figure are omitted; the limit cycles to which the system relaxes when it leaves the stationary state are shown. These figures contain two additional lines: the straight line is the load line $\mathcal{U} = \mathcal{U}_t - \mathcal{R}_s j$ and the curved line is the current–voltage characteristic $\mathcal{U} = \mathcal{U}(j)$ of the gas discharge. The point of intersection of the load line and the current–voltage characteristic on the phase plane indicates a stationary solution of the system.

Notably, the transition to chaos depicted in figure 3.3 exhibits a perfect period-doubling cascade with a striking similarity to simple non-linear mappings such as the well-known logistic map [51]. Bifurcations occur approximately at the values of the control parameter $\mathcal{U}_t = 18.315$, 18.782, 18.902, 18.939, 18.948, etc, and the transition to chaos at about $\mathcal{U}_t = 19$. The sequence of the ratios of the intervals between successive period-doubling bifurcations (this number is about 4.253 for the last bifurcation points indicated above) certainly converges to the first Feigenbaum constant $\delta = 4.6692\ldots$.

Figure 3.4 demonstrates Lorenz maps, obtained for regimes with applied voltages $\mathcal{U}_t = 18.2$, 18.6, 18.9 (shown in figure 3.3), and 19.5. These maps were obtained by registering successive local maxima of the discharge current $j(t)$, and then depicting $(n + 1)$st local maximum of $j(t)$ versus its nth maximum. (Very similar maps can be obtained also for the voltage $\mathcal{U}(t)$.) According to the results shown in figure 3.3, at $\mathcal{U}_t = 18.2$, 18.6, and 18.9, the calculated data points are concentrated around one, two, and four positions in figure 3.4, the number of which indicates the *period* of the corresponding attractor (period 1, 2, and 4). In the case of $\mathcal{U}_t = 19.5$, when the system oscillates completely chaotically, the data points nevertheless fall on a well-structured, nearly 1D curve. Again, in agreement with Feigenbaum's universality theory, this *unimodal* curve resembles the well-known logistic map.

In conclusion, it should be noted that the model used in this analysis can be improved in different ways, for example, by using a more detailed fluid model

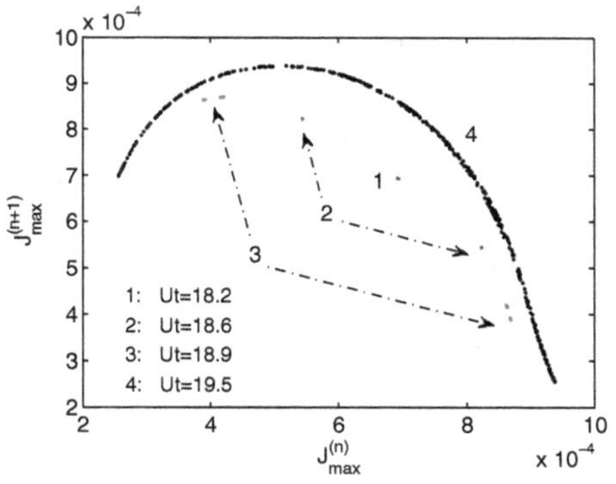

Figure 3.4. Lorenz map: $(n + 1)$st local maximum of the discharge current $j(t)$ versus its nth maximum obtained for the regimes corresponding to voltages $\mathcal{U}_t = 18.2$, 18.6, 18.9, and 19.5. The conditions are the same as in figure 3.3. Adapted from [24]. Copyright IOP Publishing. All rights reserved.

(similar to that in section 2.3.2, see also [25, 52]) and employing more realistic geometry with appropriate boundary conditions. However, it is obvious that this approximate model is already capable of correctly predicting the physical phenomenon that is being considered in this section.

3.3 Pattern formation in the gas discharge–semiconductor system

This section is devoted to the numerical study of the emergence of stationary spatially periodic patterns in the gas discharge–semiconductor system (GDSS). We focus on the experimental conditions of [10] and employ the 2D numerical model described in section 3.1, which is essentially similar to those in [23, 24, 34, 38, 39].

Calculations reveal the existence of multiple stationary patterns in GDSS arising from a stationary homogeneous mode at the same discharge current. The calculated discharge parameters are in reasonable agreement with experiment [10]. Results of calculations support the evidence [5, 33, 34] that gas heating is responsible for the instability in GDSS under the considered conditions: spatial filaments do not appear when the heat equation is excluded from the system. Therefore, the model equations described in section 3.1.1 are supplemented by the gas heating equation

$$Nc_{p_1}\frac{\partial T}{\partial t} - \alpha\nabla^2 T = \mathbf{J}_g \cdot \mathbf{E}, \tag{3.19}$$

were T is the gas temperature, N is the background gas density, α is the thermal conductivity, c_{p_1} is the heat capacity at constant pressure per molecule. On the cathode, $z = d_g$, the temperature is defined as $T = T_0 = 100$ K, on the anode, $z = 0$, we impose $\partial T/\partial z = 0$, and $\partial T/\partial x = 0$ on the lateral boundaries, $x = 0$ and $x = l$, of the discharge cell.

3.3.1 Input parameters

The input parameters correspond to the experimental conditions in [10] and are defined as in [5, 34]. A silicone semiconductor with dielectric constant of $\varepsilon_s = 11.7$, conductivity of $\sigma_s = 1.3 \times 10^{-8}$ $(\Omega\ \mathrm{cm})^{-1}$ (which can be altered through photosensitive doping) and thickness of $d_s = 1$ mm is cooled by liquid nitrogen to temperature $T_0 \approx 100$ K. The discharge gap of length $d_g = 0.44$ mm and width $l = 20$ mm is filled with nitrogen at pressure $p_1 = 211.5$ Torr. The pressure reduced to room temperature $T = 293$ K is $p = 619.8$ Torr. The secondary emission coefficient is $\gamma = 0.0026$. The ion and electron mobilities are defined as $\mu_i p = 1140\ \mathrm{cm^2\ Torr/(sV)}$ and $\mu_e p = 4.4 \times 10^5\ \mathrm{cm^2\ Torr/(sV)}$, respectively. The diffusion coefficients are $D_e = \mu_e T_e$ for electrons (with $T_e = 6$ eV) and $D_i = \mu_i T_0$ for ions. The heat conductivity is $\alpha = 0.95 \times 10^{-4}\ \mathrm{W\ cm^{-1}\ K^{-1}}$. The constants in the Townsend formula (3.6) are specified as $A = 12\ \mathrm{cm^{-1}\ Torr^{-1}}$ and $B = 342\ \mathrm{V\ cm^{-1}\ Torr^{-1}}$ [40].

3.3.2 Multiple stationary patterns

Numerical analysis reveals the emergence of different stationary spatially periodical patterns under the same discharge conditions. Figures 3.5, 3.6, and 3.7 show profiles

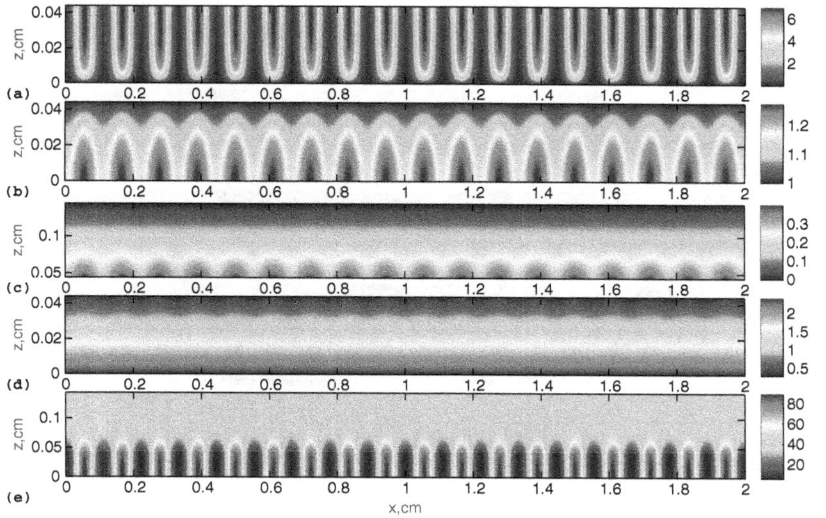

Figure 3.5. Distributions of the ion density n_i (in units of 10^9 cm^{-3}) (a) and the gas temperature T (in units of $T_0 = 100$ K) (b) in the gas discharge region, the electric potential φ (in kV) in the semiconductor region (c) and in the gas discharge region (d), the longitudinal component of the current density J_z (in $\mu A\ cm^{-2}$) in the entire gas discharge–semiconductor system (e). Applied voltage $U_0 = 2440$ V. Mode 1. Adapted from [36] with permission of AIP Publishing.

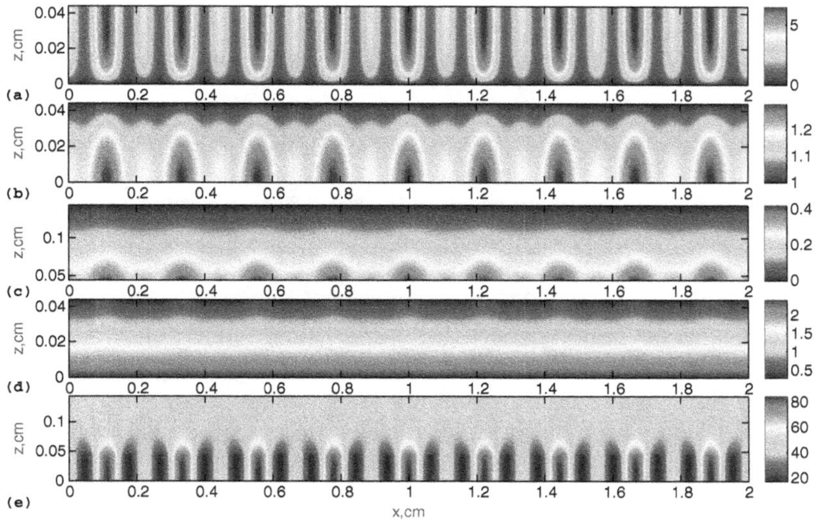

Figure 3.6. Mode 2. The conditions are the same as in figure 3.5. Adapted from [36], with permission of AIP Publishing.

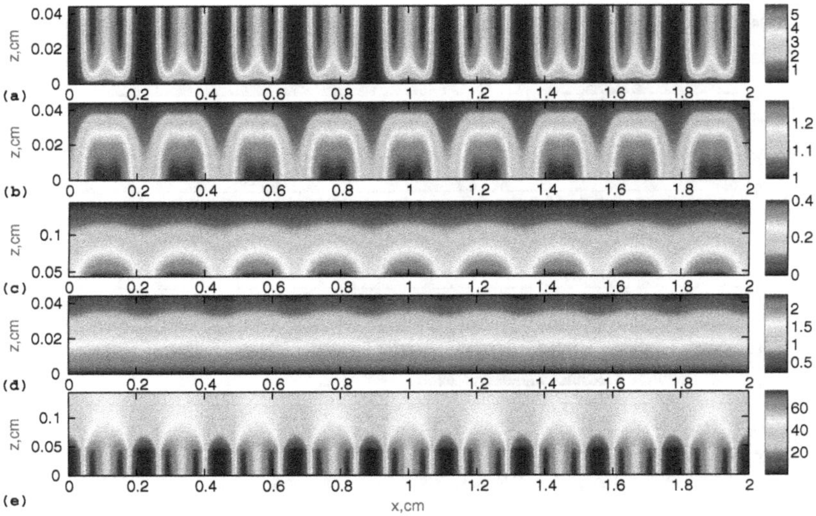

Figure 3.7. Mode 3. The conditions are the same as in figure 3.5. Adapted from [36] with permission of AIP Publishing.

of the ion density n_i (a), the gas temperature T (b), the electric potential φ in the gas discharge (c) and semiconductor (d) regions, and the longitudinal component of the current density J_z (e) in the entire gas discharge–semiconductor region, corresponding to three filamentary modes, obtained at the same value of the applied voltage $U_0 = 2440$ V.

The distributions of these parameters along the gas discharge–semiconductor interface, corresponding to the filamentary modes shown in figures 3.5–3.7, are given in figure 3.8.

The discharge currents corresponding to filamentary modes (shown in figures 3.5–3.8) and a homogeneous mode at $U_0 = 2440$ V are nearly the same (see figure 3.9); the current density for the latter is 45 μA cm^{-2}. Figure 3.9 shows the dependence of the applied voltage U_0 on the averaged (over the transverse direction x) value of the current density J for three filamentary modes and a homogeneous mode. It can be seen that filamentary mode 1 exists for applied voltages in the range 2325–2464 V (line AF in figure 3.9), mode 2 for voltages between 2421 and 2451 V (line CD), and mode 3 for the range between 2397 and 2455 V (line BE). The rectangular mark in figure 3.9 indicates the region, where the discharge parameters shown in figures 3.5–3.8 were calculated.

Figure 3.9 also illustrates the situation (see smaller panel), where the current density J is plotted against the value of the electric potential averaged over the gas discharge–semiconductor interface (the current–voltage characteristic, CVC). Note that the CVC slope is negative as it must be [5, 40].

3.3.3 Comparison of computed and experimental results

The threshold voltage $U_{0t} \sim 2420$ V for the generation of current filaments (hexagonal patterns) observed in the experiment [10] (with the semiconductor

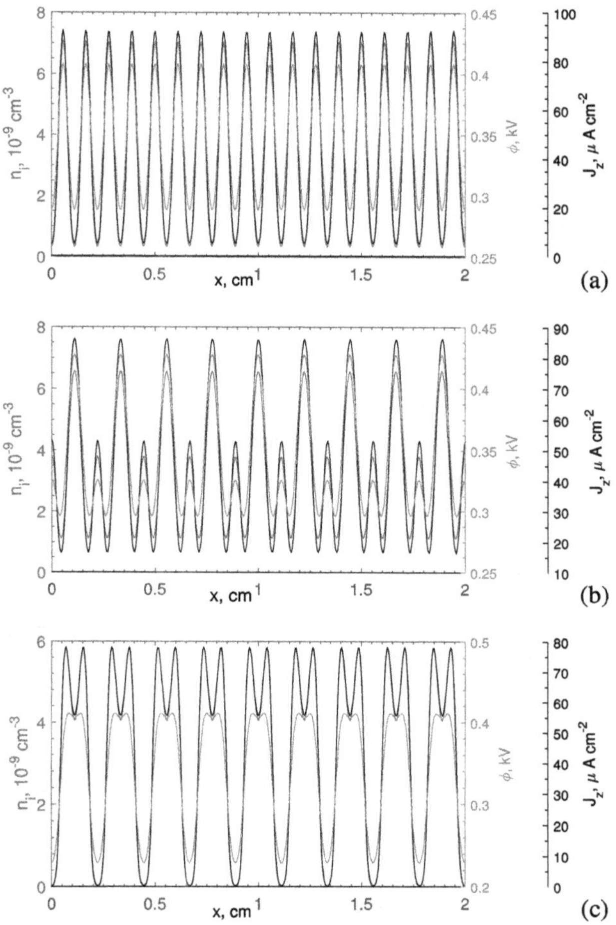

Figure 3.8. Distributions of the ion density n_i (blue), the longitudinal component of the current density J_z (black), and the electric potential (red) along the gas discharge–semiconductor interface. The conditions are the same as in figure 3.5. Mode 1 (a), mode 2 (b), mode 3 (c). Adapted from [36] with permission of AIP Publishing.

conductivity $\sigma_s = 1.3 \times 10^{-8}\ \Omega^{-1}\ \text{cm}^{-1}$) is slightly higher than that of the calculated 2325 V (see figure 3.9). The measured value of the average current density at which the patterns appear $J \sim 10\ \mu\text{A cm}^{-2}$ is correspondingly higher [10, 34] (see figure 3.9). The model that was used is a relatively simple fluid model, so that the calculated and measured results can be considered to be in reasonable agreement. In fact, one of the main sources of uncertainty in fluid models of glow discharges is the secondary electron emission coefficient γ. It depends non-linearly on the reduced electric field, E/p, and also depends on the cathode material and even on the state of the cathode surface. The secondary electron emission coefficient can be used as a fitting parameter for good agreement between calculated and measured data. In [36],

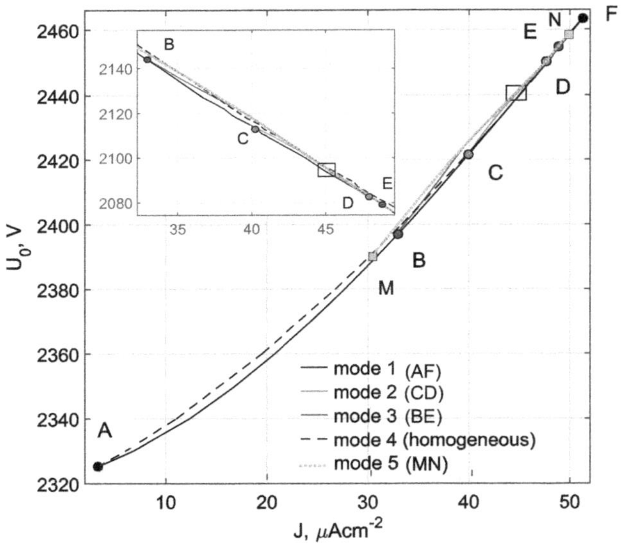

Figure 3.9. Dependence of the applied voltage U_0 on the average (over transverse x direction) current density J for different modes. The smaller panel shows the data between points B and E, with the vertical axis expressing the electric potential averaged over the gas discharge–semiconductor boundary (the current–voltage characteristic of the discharge). The circles indicate the bifurcation points. The rectangular mark indicates the region, where the discharge parameters in figures 3.5–3.8 were computed.

in order to maintain consistency with the results from [34], the same value, $\gamma = 0.0026$, was used for secondary electron emission.

In [34], under similar discharge conditions, it was reported that the stationary spatially periodic current filaments remained practically the same regardless of the initial conditions, more precisely, regardless of whether the initial conditions are purely homogeneous or perturbed by the addition of random noise. (In the first case, round-off errors due to the numerical integration acted as a perturbation.) This certainly contradicts the present results, in which different stationary modes are found. In fact, this study follows a different approach. The initial condition for the time-dependent problem was a stationary solution homogeneous in the transverse direction x (i.e., the solution corresponding to the current–voltage curve indicated as mode 4 in figure 3.9) plus a small perturbation in the x direction. Eventually, as the time-dependent solution approached a steady state, a stationary solver was launched, leading to a spatially periodic stationary pattern.

It should be noted that the spatial period of the stationary pattern, indicated here as mode 1, is exactly equal to the period observed in experiment [10], which is 0.11 cm, and which implies that the corresponding number of filaments in the transversal direction is 18 (see figures 3.5 and 3.8(a)). In [34], under similar conditions, the spatial period was found to be 0.16 cm at $U_0 = 2400$ V, which means 12.5 filaments. According to the present results, at $U_0 = 2400$ V, the number of filaments is 9 or 18 (see figure 3.9). In [34] it was also mentioned that the period became 0.25 cm at $U_0 = 2345$ V, which counts as 8 filaments, and which is 18 in our case (figure 3.9). Moreover, present calculations have shown that with a change in the applied voltage U_0 (more precisely, with sliding along one of the CVC in figure 3.9) the amplitudes of the discharge parameters change, but the number of filaments remains constant, and, therefore, what was actually observed in [34] at $U_0 = 2345$ V should be attributed to a different mode. A possible reason for the disagreement in the filament period is that stationary modes different from those observed here, were found in [34]. The difference in details of the numerical models also should be taken into account. Certainly, stationary spatially inhomogeneous modes are not limited by those described here and in [34], and only one of the possible stationary modes was observed experimentally and reported in [10].

3.3.4 Linear stability analysis

The stability analysis of a homogeneous stationary state is carried out in essentially the same way as in [39], using the ansatz

$$n_e(x, z, t) = n_{e0}(z) + n_{e1}(z)\, e^{ikx+\lambda t}, \tag{3.20}$$

$$n_i(x, z, t) = n_{i0}(z) + n_{i1}(z)\, e^{ikx+\lambda t}, \tag{3.21}$$

$$\varphi(x, z, t) = \varphi_0(z) + \varphi_1(z)\, e^{ikx+\lambda t}, \tag{3.22}$$

$$T(x, z, t) = T_0(z) + T_1(z)\, e^{ikx+\lambda t}. \tag{3.23}$$

Here, k denotes the wavenumber of the transverse Fourier mode, and λ is the eigenvalue. The subscript 0 indicates a stationary solution that is homogeneous in the transverse direction, and subscript 1 the amplitude of the small perturbation about this solution. The stationary solution is obtained from the solution of the system of equations (3.1)–(3.8), (3.19) subject to the boundary conditions (3.9)–(3.13), under the condition that the solution is confined to the longitudinal z-axis.

Substitution of the ansatz in the gas discharge equations (3.1)–(3.8), (3.19) and boundary conditions (3.9)–(3.13) yields

$$\begin{aligned}\lambda n_{e1} - \frac{d}{dz}\left(D_e\frac{dn_{e1}}{dz} + \mu_e n_{e1}E_0\right) + k^2 D_e n_{e1}\\ - \mu_e\left(\frac{dn_{e0}}{dz}E_{z1} + n_{e0}\frac{dE_{z1}}{dz} + k^2 n_{e0}\varphi_1\right) = S_1,\end{aligned} \tag{3.24}$$

$$\begin{aligned}\lambda n_{i1} - \frac{d}{dz}\left(D_i\frac{dn_{i1}}{dz} - \mu_i n_{i1}E_0\right) + k^2 D_i n_{i1}\\ + \mu_i\left(\frac{dn_{i0}}{dz}E_{z1} + n_{i0}\frac{dE_{z1}}{dz} + k^2 n_{i0}\varphi_1\right) = S_1,\end{aligned} \tag{3.25}$$

$$-\frac{d^2\varphi_1}{dz^2} + k^2\varphi_1 = \frac{e}{\varepsilon_0}(n_{i1} - n_{e1}), \tag{3.26}$$

$$\lambda N c_{p_1} T_1 - \alpha \frac{d^2 T_1}{dz^2} + k^2 \alpha T_1 = S_{T1}. \tag{3.27}$$

Here, $E_0 = -d\varphi_0/dz$ is the unperturbed electric field and $E_{z1} = -d\varphi_1/dz$ is z component of the electric field perturbation. The right-hand sides S_1 and S_{T1} of the equations (3.24), (3.25), and (3.27) represent perturbations of the source terms of the continuity equations for electrons and ions (3.1), (3.2) and of the heat equation (3.19).

The boundary conditions at anode $z = 0$ obtain the form

$$n_{i1} = 0, \quad \frac{dn_{e1}}{dz} = 0, \quad \frac{dT_1}{dz} = 0, \quad \varphi_1 = 0. \tag{3.28}$$

On the interface between the gas discharge and semiconductor layers, $z = d_g$,

$$\frac{dn_{i1}}{dz} = 0, \quad \mu_e n_{e1} = \gamma \mu_i n_{i1}, \quad T_1 = 0. \tag{3.29}$$

The solution of the Poisson equation in the semiconductor region, $d_g < z < d_g + d_s$, can be found explicitly:

$$\varphi_1(z) = C \sinh[k(z - d_g - d_s)].$$

Using this solution and excluding the perturbed surface charge density Q_1 from equations (3.11) and (3.12), we obtain

$$-Ck \cosh(kd_s)(\lambda\varepsilon_0\varepsilon_s + \sigma_s) = (J_{gz1} + \lambda\varepsilon_0 E_{z1})|_{d_g^-},$$

from which the constant C is determined. Finally, due to continuity of the electric potential on the interface between the gas discharge and semiconductor, $\varphi_1(d_g^-) = \varphi_1(d_g^+)$, the second boundary condition for the potential is obtained in the form

$$\varphi_1(d_g) = \left.\frac{J_{gz1} + \lambda\varepsilon_0 E_{z1}}{\lambda\varepsilon_0\varepsilon_s + \sigma_s}\right|_{d_g^-} \frac{\tanh(kd_s)}{k}. \tag{3.30}$$

Ordinary differential equations (3.24)–(3.27) together with boundary conditions (3.28)–(3.30) form an eigenvalue problem for $\lambda(k)$.

Here we carry out the linear stability analysis of the homogeneous stationary state, which is specified as mode 4 in figure 3.9. By this method, the system of partial differential equations (3.1)–(3.8), (3.19) and boundary conditions (3.9)–(3.13) describing the GDSS is reduced to the eigenvalue problem (3.24)–(3.30), which is a boundary-value problem for ordinary differential equations. By solving this problem for a specific set of parameters, a dispersion relation $\lambda = \lambda(k)$ is derived that can be used to predict the behavior of the GDSS for a given parameter regime.

Indeed, the most unstable mode, which is mode $k = k^*$ with the greatest positive growth rate $\mathrm{Re}(\lambda)$, dominates the dynamical behavior of the GDSS with respect to the stationary state. If $\mathrm{Re}(\lambda)$ is negative for all k, the system is dynamically stable. If patterns emerge spontaneously in the system due to linear instability, there will be a band of modes with positive growth rate $\mathrm{Re}(\lambda(k)) > 0$. If the instability simply grows or shrinks without oscillations, then $\mathrm{Im}(\lambda(k)) = 0$, and, in dynamical terms, this case is classified as Turing instability. On the other hand, if the imaginary part of the dispersion relation does not vanish, $\mathrm{Im}(\lambda(k)) \neq 0$, then the system oscillates. If the system oscillates in such a way that the most unstable mode $k = k^*$ has no spatial structure, this implies that $k^* = 0$ and we speak of a Hopf transition, whereas if $k^* \neq 0$, the transition is called Turing–Hopf.

The dispersion curves obtained in the case when the heat conduction equation (3.19) and, accordingly, equation (3.27) are excluded from the consideration, are shown in figure 3.10. (In this case, we defined $T \equiv T_0 = 100\ K$.) More precisely, this figure represents the dependence of the maximum of $\mathrm{Re}(\lambda(k))$ on the wavenumber k for several values of the applied voltage ($U_0 = 2400$, 2440, and 2480 V at $p_1 = 211.5$ Torr) and pressure ($p_1 = 206.2$ and 222.1 Torr at $U_0 = 2440$ V). Calculations have shown that the dependence of the dispersion curve on the applied voltage U_0 is rather weak. Dependence of max $(\mathrm{Re}\,(\lambda(k)))$ on the conductivity σ_s also appeared to be weak; we examined for the values of σ_s from the experiment [10], namely, for $\sigma_s = 1.3 \times 10^{-8}$, 4.0×10^{-9}, and 6.3×10^{-9} $(\Omega\ \mathrm{cm})^{-1}$ (the corresponding dispersion curves are not shown in the figure).

As can be seen from figure 3.10, the most unstable mode is always such that $k^* = 0$ and $\mathrm{Re}(\lambda(k^*)) > 0$. Moreover, eigenvalues λ usually appear in complex conjugate pairs when solving the eigenvalue problem (3.24)–(3.26), (3.28)–(3.30).

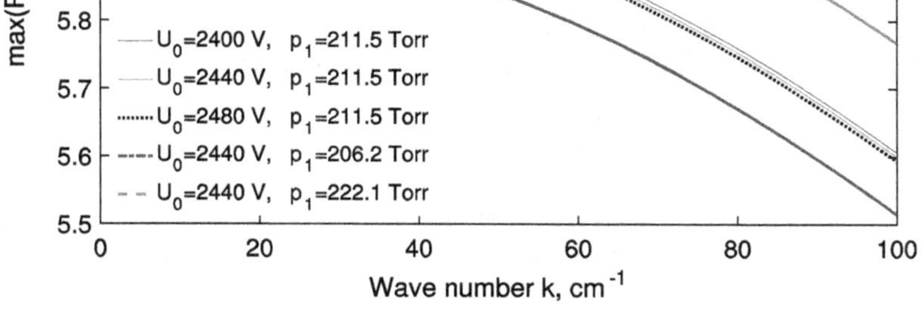

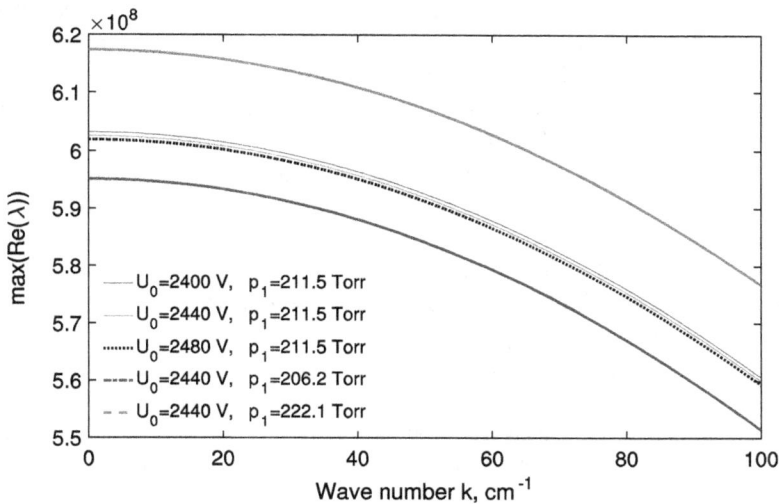

Figure 3.10. The real parts of the dispersion relation $\lambda(k)$. The gas heating is not taken into account. Adapted from [36] with permission of AIP Publishing.

However, it was found that $\mathrm{Im}(\lambda(k)) = 0$ for the eigenvalues with maximum real parts, that is, for $\lambda(k)$, which form the dispersion curves in figure 3.10, and, therefore, $\mathrm{Im}(\lambda(k^*)) = 0$. This implies that in the considered parameter regime, the homogeneous stationary state for GDSS is unstable and that the instability does not form a spatial structure: it is homogeneous and grows without oscillations.

This certainly agrees with the results of the direct numerical solution of the system (3.1)–(3.8), (3.9)–(3.14), without taking into account gas heating. In fact, when the heat conduction equation (3.19) is excluded from the model and the gas temperature is set to $T \equiv T_0 = 100\,K$, the occurrence of stationary current filaments is not observed in this case either. The same effect of gas heating was observed earlier in numerical modeling in [34] that led to the conclusion that the gas heating was responsible for the instability in GDSS under the conditions of experiment [10]. The dispersion curves change drastically when gas heating is taken into account: the dependence of the maximum of $\mathrm{Re}(\lambda(k))$ on the wavenumber becomes irregular. Although the current filaments are highly non-linear objects and develop well beyond the limits of the linear stability analysis, the present result can be considered as indirect evidence of the destabilizing effect of gas heating on GDSS.

3.3.5 Spontaneous division of current filaments

The emergence of spatially localized solitary objects known as dissipative solitons (DSs) is a distinctive feature of self-organization in spatially extended systems operating far from equilibrium. In DC gas discharge system with high-ohmic electrodes, these objects are generated in the form of current filaments as a result of Turing destabilization of the homogeneous stationary state of the system [4, 26–28].

It is important to note that the physical mechanism of instability in this system leading to filamentation of the current is fundamentally different from that in dielectric barrier discharges (DBD) (see [5] for details). Indeed, the effect of space charge redistribution of the longitudinal electric field, which usually causes instability in DBD (and, more generally, in glow-like discharges), is negligible in GDSS under cryogenic conditions due to low current density [5, 33, 34]. The thermal mechanism was proposed in [33] and further corroborated in [34–36] to explain the filamentation of the current in GDSS. As was mentioned in the previous section, the calculations based on a truncated model, obtained by excluding the gas heating equation (3.19) (i.e., assuming a constant gas temperature), do not lead to instability and filamentation [5, 34–36].

Experiments evidence that one of the possible mechanisms of DSs generation is the division of current filaments [18]. More specifically, three different scenarios for DSs division were identified in [18]. The first is related to DSs splitting as they travel in the device at constant values of control parameters. The second scenario consists of the splitting process of a single quasi-stationary radially symmetric DS with increasing the driving voltage U_0. The third scenario corresponds to the process of division for a pattern of spatially dense arrangement of DSs (see figure 3.11). The original pattern represents a spatially periodic distribution of identical electric current filaments in the active area of the device (figure 3.11(a)). With increase in the discharge current (an increase in the semiconductor conductivity by its irradiation with IR light), the DS amplitude increases and the spatial period of the pattern grows (figure 3.11(b, c)). At a certain stage in the growth of the DC pattern, they begin to divide (figure 3.11(d)) [18].

We focus on the third scenario for DSs division. The GDSS under the conditions in section 3.3.1 that were previously used in [36] (and also in [5, 34, 35]) is considered. The multiple stationary filamentary modes for GDSS, coexisting in the same current range [36], were described in the previous section. The current–voltage curves corresponding to those modes are denoted as modes 1–4 in figure 3.9. This figure shows the profiles of the applied voltage U_0 as functions of the average (over the transverse direction x) discharge current density J. The additional mode 5 in figure 3.9 that was detected in [53] is directly related to the DSs division. Indeed, numerical analysis reveals the division process for a pattern of spatially dense arrangement of stationary DSs, similar to that observed in experiment [18], in the relevant parameter regime. Figure 3.12 demonstrates the growth of the current filaments until they reach a critical size and the subsequent splitting of these filaments as the discharge current is increased with an increase in the supply voltage U_0 = 2400, 2428, 2435, and 2455 V.

Figures 3.13–3.16 demonstrate the profiles of the ion density n_i (a), the gas temperature T (b), the electric potential φ (c) and (d), and the longitudinal component of the current density J_z (e), corresponding to four states in figure 3.12.

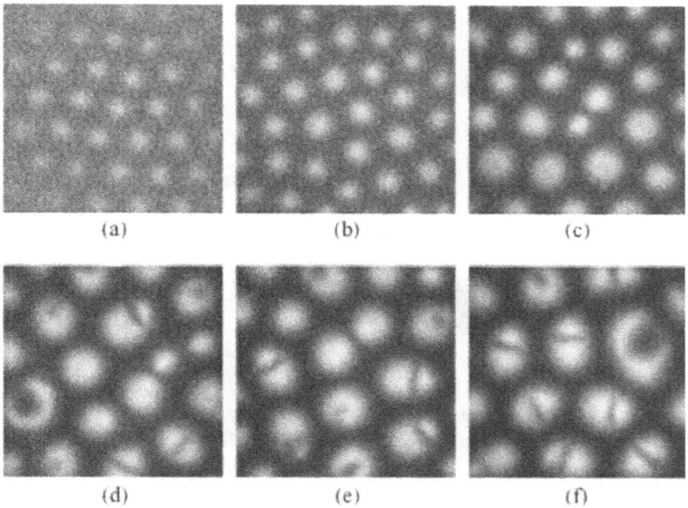

Figure 3.11. Snapshots of the transition process from small to large amplitude pattern and subsequent division phenomenon. Discharge gap d_g = 1 mm, pressure 200 hPa, applied voltage U_0 = 2.8 kV. The patterns (a)–(f) correspond to the increase of the averaged current density over the range 1.1–5.2 μA cm^{-2}, in response to the increase of intensity of IR light illuminating the semiconductor. Adapted with permission from [18]. Copyright 2006 Elsevier.

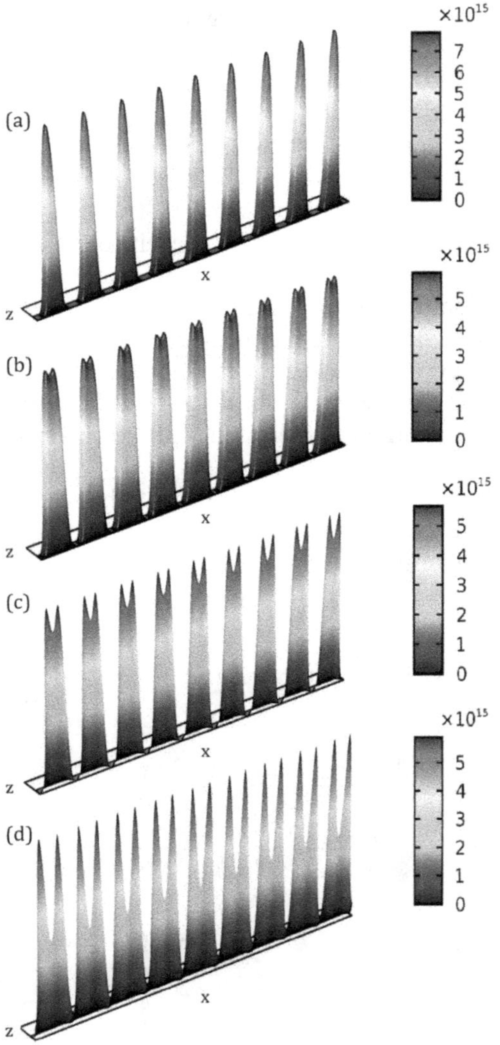

Figure 3.12. Spontaneous splitting of DSs with increase in the discharge current. Spatial distribution of the ion density n_i (in units of m^{-3}) at the applied voltage (a) $U_0 = 2400$ V, (b) 2428 V, (c) 2435 V, and (d) 2455 V. Discharge gap $d_g = 0.44$ mm, pressure $p_1 = 211.5$ Torr. The states (a)–(d) correspond to the mode 5 (curve MN) in figure 3.9. Adapted from [53] with permission of AIP Publishing.

A strong qualitative agreement is evident between the experimentally observed process of filament division (figure 3.11) and the modeling results (figures 3.12–3.16). However, it is important to note that the actual dynamics of DSs in GDSS is quite complicated and not limited to the scenarios described in experiment [18]. For example, the numerical model identifies not only the growth of additional filaments with increase in the discharge current, but it also

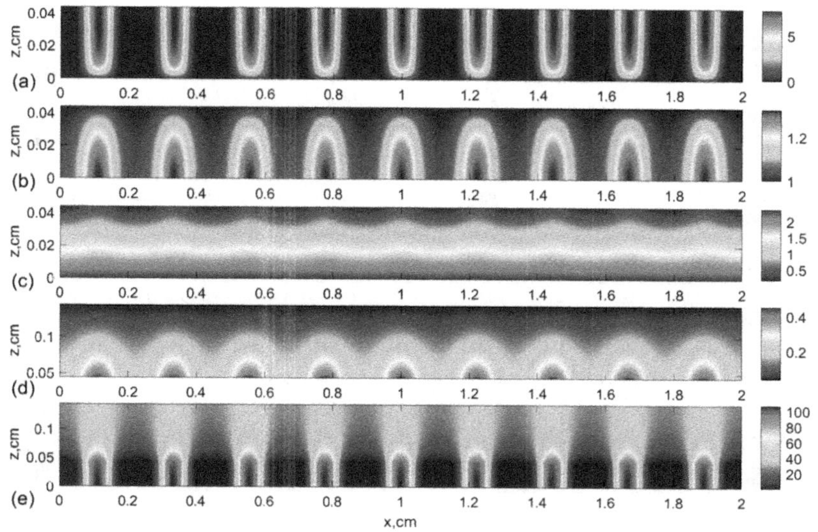

Figure 3.13. Spatial profiles of the ion density n_i (in units of 10^9 cm^{-3}) (a) and the temperature T (in units of T_0 = 100 K) (b) in the gas discharge part of the GDSS, the electric potential φ (in kV) in the semiconductor (c) and the gas discharge parts (d), the component J_z of the current density (in μA cm^{-2}) in the entire GDSS (e). Applied voltage U_0 = 2400 V corresponds to state (a) in figure 3.12. Adapted from [53] with permission of AIP Publishing.

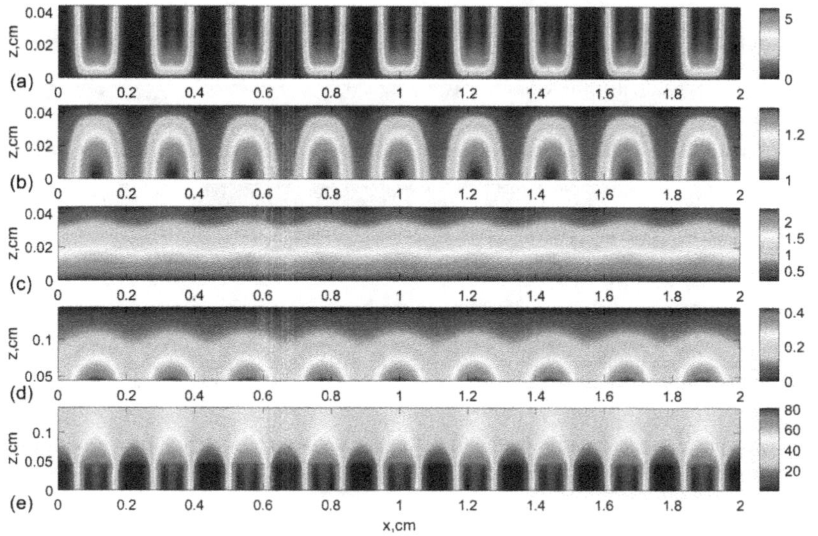

Figure 3.14. The same parameters as in figure 3.13. Applied voltage U_0 = 2428 V corresponds to state (b) in figure 3.12. Adapted from [53] with permission of AIP Publishing.

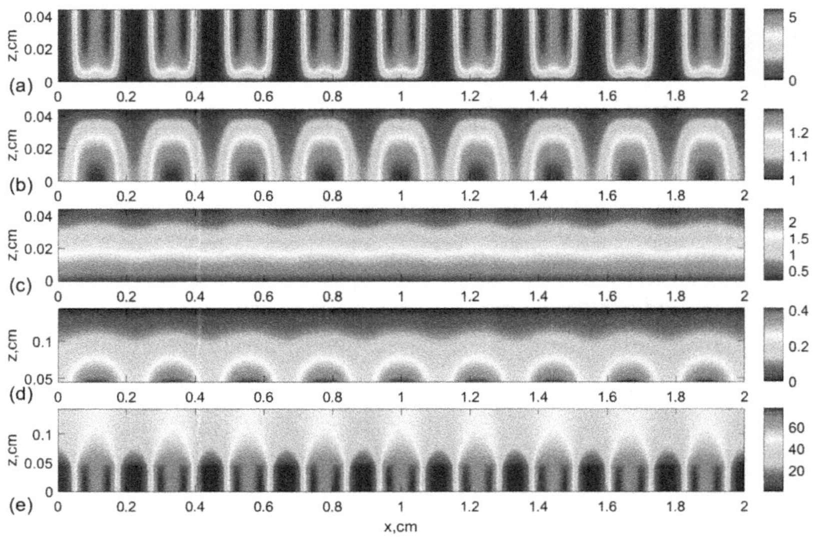

Figure 3.15. The same parameters as in figure 3.13. Applied voltage $U_0 = 2435$ V corresponds to state (c) in figure 3.12. Adapted from [53] with permission of AIP Publishing.

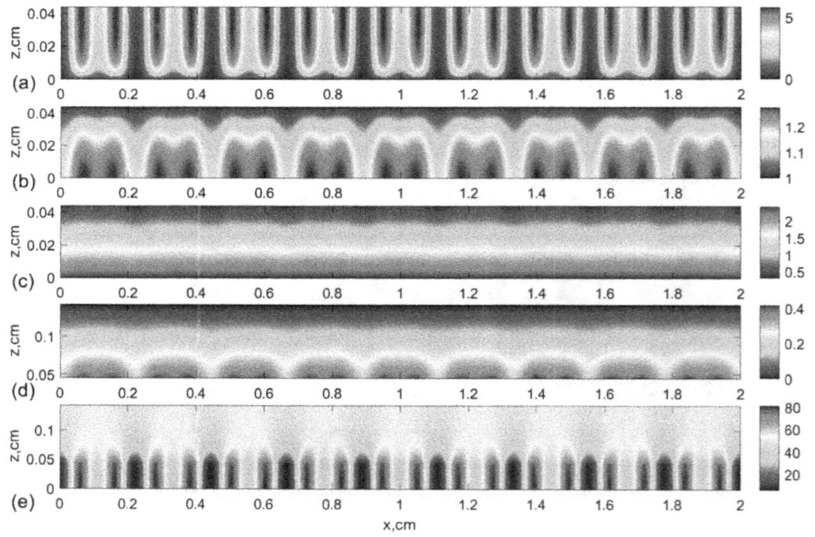

Figure 3.16. The same parameters as in figure 3.13. Applied voltage $U_0 = 2455$ V corresponds to state (d) in figure 3.12. Adapted from [53] with permission of AIP Publishing.

reveals the extinction of these filaments. This behavior, as shown in figure 3.17, is associated with mode 3.

It should be noted that filamentary modes other than mode 5 may also exhibit spontaneous filament division as the control parameters are changed. An example is shown in figure 3.18 for mode 1, where a filament splitting process is observed with a smooth increase in the semiconductor conductivity from $\sigma_s = 1.3 \times 10^{-8}$ $(\Omega\ \text{cm})^{-1}$ to $5 \times 1.3 \times 10^{-8}$ $(\Omega\ \text{cm})^{-1}$, while maintaining a constant voltage $U_0 = 2320$ V.

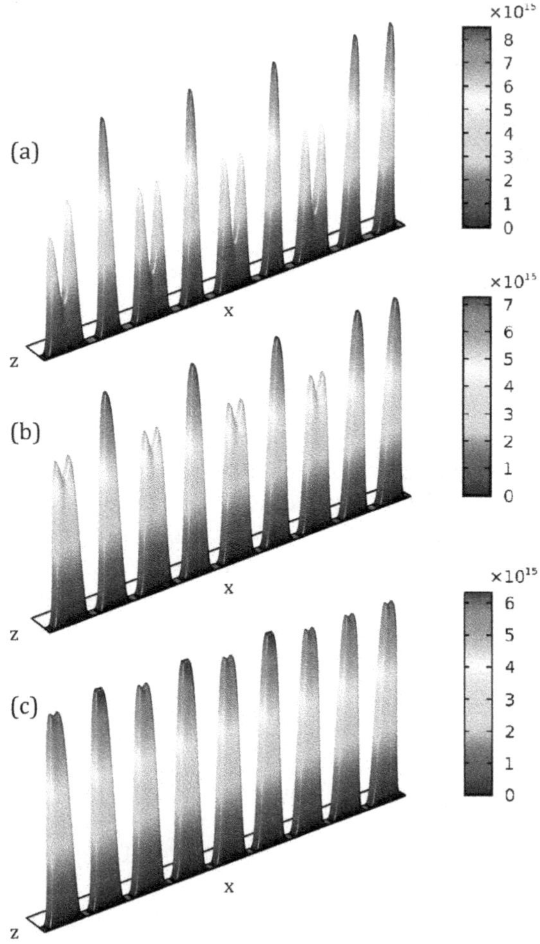

Figure 3.17. Extinction of additional filaments with increase in the discharge current. Spatial distribution of the ion density n_i (in units of m^{-3}) at the applied voltage (a) $U_0 = 2397$ V, (b) 2421 V, and (c) 2425 V. The remaining conditions are the same as in figure 3.12. The states (a)–(c) correspond to the mode 3 (curve BE) in figure 3.9. Adapted from [53] with permission of AIP Publishing.

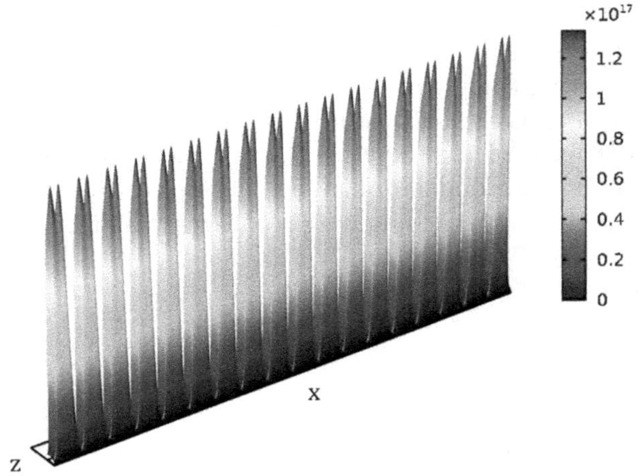

Figure 3.18. The onset of filament division with increase in the conductivity of the semiconductor cathode from $\sigma_s = 1.3 \times 10^{-8}\,(\Omega\,\mathrm{cm})^{-1}$ to $5 \times 1.3 \times 10^{-8}\,(\Omega\,\mathrm{cm})^{-1}$. Spatial distribution of the ion density n_i (in units of m^{-3}) at $U_0 = 2320$ V. The remaining conditions are the same as in figure 3.12. An *unperturbed* state (corresponding to $\sigma_s = 1.3 \times 10^{-8}\,(\Omega\,\mathrm{cm})^{-1}$) belongs to the mode 1 (curve AF) in figure 3.9. Adapted from [53] with permission of AIP Publishing.

In conclusion, we emphasize that the results presented in this section were obtained using a stationary solver. More precisely, the time-dependent solver was used first, and when the solution was close to its stationary state, the stationary solver was launched. That is why this study is limited to the analysis of stationary DS patterns, and, therefore, is related to the third scenario of DSs division in experiment [18]. It is reasonable to predict that time-dependent numerical solutions will be able to reproduce evolutionary scenarios for division of propagating DSs that were observed in [18].

References

[1] Cross M C and Hohenberg P C 1993 Pattern formation outside of equilibrium *Rev. Mod. Phys.* **65** 851–1112

[2] Purwins H-G, Bödeker H U and Amiranashvili S 2010 Dissipative solitons *Adv. Phys.* **59** 485–701

[3] Astrov Y A 2010 Semiconductor-gas-discharge planar structures as devices for unconventional computing *Int. J. Unconv. Comput.* **6** 33–73

[4] Purwins H-G, Bödeker H U and Liehr A W 2005 *Dissipative Solitons* (Lecture Notes in Physics vol 661) ed N Akhmediev and A Ankiewicz (Berlin: Springer) p 267

[5] Raizer Y P and Mokrov M S 2013 Physical mechanisms of self-organization and formation of current patterns in gas discharges of the Townsend and glow types *Phys. Plasmas* **20** 101604

[6] Almeida P G C and Benilov M S 2013 Multiple solutions in the theory of direct current glow discharges: effect of plasma chemistry and nonlocality, different plasma-producing gases, and 3d modelling *Phys. Plasmas* **20** 101613
[7] Astrov Y A and Logvin Y A 1997 Formation of clusters of localized states in a gas discharge system via a self-completion scenario *Phys. Rev. Lett.* **79** 2983–6
[8] Ammelt E, Astrov Y A and Purwins H-G 1998 Hexagon structures in a two-dimensional dc-driven gas discharge system *Phys. Rev.* E **58** 7109–17
[9] Astrov Y A, Müller I, Ammelt E and Purwins H-G 1998 Zigzag destabilized spirals and targets *Phys. Rev. Lett.* **80** 5341–4
[10] Astrov Y A, Lodygin A N and Portsel L M 2011 Hexagonal structures of current in a 'semiconductor-gas-discharge gap' system *Tech. Phys.* **56** 197–203
[11] Strümpel C, Astrov Y A and Purwins H-G 2000 Nonlinear interaction of homogeneously oscillating domains in a planar gas discharge system *Phys. Rev.* E **62** 4889–97
[12] Strümpel C, Purwins H-G and Astrov Y A 2001 Spatiotemporal filamentary patterns in a dc-driven planar gas discharge system *Phys. Rev.* E **63** 026409
[13] Ammelt E, Astrov Y A and Purwins H-G 1997 Stripe Turing structures in a two-dimensional gas discharge system *Phys. Rev.* E **55** 6731–40
[14] Astrov Y, Ammelt E, Teperick S and Purwins H-G 1996 Hexagon and stripe Turing structures in a gas discharge system *Phys. Lett.* A **211** 184–90
[15] Gurevich E L, Astrov Y A and Purwins H G 2005 Pattern formation in planar dc-driven semiconductor–gas discharge devices: two mechanisms *J. Phys. D: Appl. Phys.* **38** 468
[16] Strümpel C, Astrov Y A and Purwins H-G 2002 Multioscillatory patterns in a hybrid semiconductor-gas-discharge system *Phys. Rev.* E **65** 066210
[17] Gurevich S V, Amiranashvili S and Purwins H-G 2006 Breathing dissipative solitons in three-component reaction-diffusion system *Phys. Rev.* E **74** 066201
[18] Astrov Y A and Purwins H-G 2006 Spontaneous division of dissipative solitons in a planar gas-discharge system with high ohmic electrode *Phys. Lett.* A **358** 404–8
[19] Portsel L M, Lodygin A N and Astrov Y A 2009 Townsend-like discharge: the suppression of instabilities by a semiconductor electrode *J. Phys. D: Appl. Phys.* **42** 235208
[20] Astrov Y A, Lodygin A N and Portsel L M 2014 Dynamics and stability of the Townsend discharge in nitrogen in narrow gaps *Phys. Rev.* E **89** 033109
[21] Purwins H G and Stollenwerk L 2014 Synergetic aspects of gas-discharge: lateral patterns in dc systems with a high ohmic barrier *Plasma Phys. Control. Fusion* **56** 123001
[22] Willebrand H, Hünteler T, Niedernostheide F-J, Dohmen R and Purwins H-G 1992 Periodic and turbulent behavior of solitary structures in distributed active media *Phys. Rev.* A **45** 8766–75
[23] Šijačić D D, Ebert U and Rafatov I 2004 Period doubling cascade in glow discharges: local versus global differential conductivity *Phys. Rev.* E **70** 056220
[24] Rafatov I 2016 Three-dimensional numerical modelling of temporal and spatial pattern formation in a dc-driven gas discharge-semiconductor system *Plasma Sources Sci. Technol.* **25** 065014
[25] Rafatov I and Yesil C 2018 Transition from homogeneous stationary to oscillating state in planar gas discharge-semiconductor system in nitrogen: effect of fluid modelling approach *Phys. Plasmas* **25** 082107
[26] Bode M and Purwins H-G 1995 Pattern formation in reaction–diffusion systems—dissipative solitons in physical systems *Physica* D **86** 53–63

[27] Kerner B S and Osipov V V 2013 *Autosolitons: A New Approach to Problems of Self-Organization and Turbulence* vol 61 (Berlin: Springer Science & Business Media)
[28] Liehr A 2013 *Dissipative Solitons in Reaction Diffusion Systems* vol 70 (Berlin: Springer)
[29] Astrov Y A and Purwins H-G 2001 Plasma spots in a gas discharge system: birth, scattering and formation of molecules *Phys. Lett.* A **283** 349–54
[30] Bode M, Liehr A W, Schenk C P and Purwins H-G 2002 Interaction of dissipative solitons: particle-like behaviour of localized structures in a three-component reaction-diffusion system *Physica* D **161** 45–66
[31] Pearson J E 1993 Complex patterns in a simple system *Science* **261** 189–92
[32] Niedernostheide F-J, Dohmen R, Willebrand H, Schulze H-J and Purwins H-G 1992 Pattern formation in nonlinear physical systems with characteristic electric properties *Nonlinearity with Disorder* (Berlin: Springer) pp 282–309
[33] Raizer Y P and Mokrov M S 2010 A simple physical model of hexagonal patterns in a Townsend discharge with a semiconductor cathode *J. Phys. D: Appl. Phys.* **43** 255204
[34] Mokrov M S and Raizer Y P 2011 Simulation of current filamentation in a dc-driven planar gas discharge–semiconductor system *J. Phys. D: Appl. Phys.* **44** 425202
[35] Mokrov M S and Raizer Y P 2018 3d simulation of hexagonal current pattern formation in a dc-driven gas discharge gap with a semiconductor cathode *Plasma Sources Sci. Technol.* **27** 065008
[36] Rafatov I 2016 Multiple stationary filamentary states in a planar dc-driven gas discharge-semiconductor system *Phys. Plasmas* **23** 123506
[37] Willebrand H, Niedernostheide F-J, Ammelt E, Dohmen R and Purwins H-G 1991 Spatio-temporal oscillations during filament splitting in gas discharge systems *Phys. Lett.* A **153** 437–45
[38] Šijačić D D and Rafatov I 2005 Oscillations in dc driven barrier discharges: numerical solutions, stability analysis, and phase diagram *Phys. Rev.* E **71** 066402
[39] Rafatov I R, Šijačić D D and Ebert U 2007 Spatiotemporal patterns in a dc semiconductor–gas-discharge system: stability analysis and full numerical solutions *Phys. Rev.* E **76** 036206
[40] Raizer Y P 1991 *Gas Discharge Physics* (Berlin: Springer)
[41] Klinger T, Schröder C, Block D, Greiner F, Piel A, Bonhomme G and Naulin V 2001 Chaos control and taming of turbulence in plasma devices *Phys. Plasmas* **8** 1961–8
[42] Manneville P 1990 *Dissipative Structures and Weak Turbulence* (San Diego, CA: Academic)
[43] Wilson R B and Podder N K 2007 Observation of period multiplication and instability in a dc glow discharge *Phys. Rev.* E **76** 046405
[44] Ott E 2002 *Chaos in Dynamical Systems* (Cambridge: Cambridge University Press)
[45] Jackson E A 1989 *Perspectives of Nonlinear Dynamics* vol 1 (Cambridge: Cambridge University Press)
[46] Jackson E A 1991 *Perspectives of Nonlinear Dynamics* vol 2 (Cambridge: Cambridge University Press)
[47] Braun T, Lisboa J A, Francke R E and Gallas J A C 1987 Observation of deterministic chaos in electrical discharges in gases *Phys. Rev. Lett.* **59** 613–6
[48] Pugliese E, Meucci R, Euzzor S, Freire J G and Gallas J A C 2015 Complex dynamics of a dc glow discharge tube: experimental modeling and stability diagrams *Sci. Rep.* **5** 8447
[49] Mansuroglu D, Uzun-Kaymak I U and Rafatov I 2017 An evidence of period doubling bifurcation in a dc driven semiconductor-gas discharge plasma *Phys. Plasmas* **24** 053503

[50] Feigenbaum M J 1978 Quantitative universality for a class of nonlinear transformations *J. Stat. Phys.* **19** 25–52

[51] May R M 1976 Simple mathematical models with very complicated dynamics *Nature* **261** 459–67

[52] Yuan C, Yesil C, Yao J, Zhou Z and Rafatov I 2020 Transition from periodic to chaotic oscillations in a planar gas discharge–semiconductor system *Plasma Sources Sci. Technol.* **92** 065009

[53] Rafatov I 2019 Numerical evidence of spontaneous division of dissipative solitons in a planar gas discharge-semiconductor system *Phys. Plasmas* **26** 092105

www.ingramcontent.com/pod-product-compliance
Lightning Source LLC
Chambersburg PA
CBHW082104210326
41599CB00033B/6579